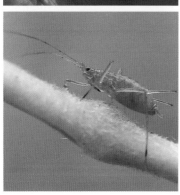

Common Insects
of Nunavut

Carolyn Mallory

Published in Canada by Inhabit Media Inc.
www.inhabitmedia.com

Inhabit Media Inc. (Iqaluit), P.O. Box 11125, Iqaluit, Nunavut, X0A 1H0
(Toronto), 146A Orchard View Blvd., Toronto, Ontario, M4R 1C3

Design and layout copyright © 2012 Inhabit Media Inc.
Text copyright © 2012 by Carolyn Mallory
Cover photo © W Lynch/Arcticphoto

Printed in Canada

All rights reserved. The use of any part of this publication reproduced, transmitted in any form or by any means, electronic, mechanical, photocopying, recording, or otherwise, or stored in a retrievable system, without written consent of the publisher, is an infringement of copyright law.

We acknowledge the support of the Canada Council for the Arts for our publishing program.

Library and Archives Canada Cataloguing in Publication

Mallory, Carolyn
    Common insects of Nunavut / Carolyn Mallory.

Includes bibliographical references and index.
ISBN 978-1-927095-00-3

1. Insects--Nunavut. 2. Insects--Nunavut--
Identification. 3 Traditional ecological knowledge--
Nunavut. I. Title. QL491.M35 2012   595.709719'5   C2011-905818-9

 Canada Council   Conseil des Arts
for the Arts   du Canada

Department of Education
Department of Culture, Language, Elders and Youth
Department of Environment
Canada Council for the Arts

Common Insects
of Nunavut

Carolyn Mallory

INHABIT
MEDIA

Credits

This book could not have been completed without the generosity of the **Nunavut Department of Environment**, who paid for the author to attend a two-week field course on Arctic and boreal insects in Churchill, Manitoba. The **Nunavut Wildlife Management Board** provided financial assistance and guidance throughout the course of this project. The **Department of Education** helped enormously by lending **Gwen Frankton** to the project and by providing financial support. **Parks Canada** helped fund the collection of traditional knowledge from across Nunavut. The **Department of Culture, Language, Elders and Youth**; the **Canadian Wildlife Service**; and **Heritage Canada** generously provided financial assistance during the production of the book.

Photographs were graciously donated by all photographers in the book, and it is a much richer product as a result. Thanks to **Johannes F. Skaftason, Susan Aiken, Richard Bartz, Valerie Behan-Pelletier, Thomas Bentley, Peter Bryant, Gwen Frankton, Henri Goulet, Ron Hemberger, Mary Hopson, Andre Karwath, James Lindsey, Jorgen Lissner, Doug Macaulay, Carolyn Mallory, Olivia Mallory, Tom Murray, Scott Nelson, P.G. Penketh, Floyd Schrock, Stefan Sollfors, Bill Stark, Kathy Thornhill, Bev Wigney,** and **Anna Ziegler**.

Danny Christopher provided the wonderful illustrations.

Grateful thanks to **Neil Christopher**, who organized an opportunity for some NTEP students to go into the communities to talk to elders about insects. Thanks to the students: **Brenda Qiyuk, Rebecca Hainnu, Sylvia Inuaraq,** and **Ellen Ittunga**. And thanks to **Annie Kellogok** for helping to collect information in Kugluktuk.

Many elders provided their knowledge on insects, and in particular I thank: **Silas Aittauq, Miriam Qiyuk, Norman Attungala, Rita Oosuaq, Aulaqiaq Areak, Kalluk Palituq, Lydia Jaypoody, Peter Kunilusie, Nellie Hikok Kanovak, Alice Ayalik, Joseph Niptanatiak, Bessie Hayokhok, Celestin Erkidjuk, Marie Ell, Neevee Nowdlak, Pauline Erkidjuk, Mary Ittunga, Judas Ikadliyuk, Bernadette Uttaq,** and **Lucy Ikadliyuk**.

I have benefitted greatly from the assistance and support of: **Rick Armstrong, Neil Christopher, Mathieu Dumond, Gwen Frankton, Henri Goulet, Peter Kevin, Mark Mallory, Jim Noble, Doug and Laurie Post, Tony Romito, Rob Roughley, Cory Sheffield, Johannes F. Skaftason, Anna Ziegler,** and **Ruth Devries**.

Acknowledgements

A work of this magnitude cannot be done without a lot of help. I thank **Mark Mallory** for reading every section of the book more than once and providing invaluable feedback. Your support means the world to me.

I also thank **Peter Kevin** and **Rob Roughley** for believing that this was a worthwhile project, teaching me about insects, and providing me with guidance and comments. Their love of insects and generosity in sharing their knowledge was not only inspirational but contagious.

A very big thank you to **Johannes F. Skaftason** for his wonderful passion for insect photography. We met online and then in person in Iceland. He has been unbelievably supportive and without his generous photographic contributions, this book would not be nearly as beautiful.

Thanks to **Conor**, **Jessamyn**, and **Olivia** for making me smile throughout the process. You are the best!

Dedication

I dedicate this book to **Dr. Rob Roughley**, who died unexpectedly before the book went to print. Rob helped me to clarify what the book should include at its very inception. Learning about insects from Rob was both a lot of fun and a real privilege. He always made time for me—no question was too foolish.

Biography

Carolyn Mallory lives in Iqaluit, where she divides her time between library work and writing. Since moving to Nunavut in 1999, she has dedicated much of her time to learning about the natural world around her. This is her second book, following the very popular *Common Plants of Nunavut*, which she co-wrote with Susan Aiken. Insects seemed like a natural follow-up to plants.

Carolyn is currently working on several fiction projects and had her first poem published this year.

Johannes F. Skaftason was born in Iceland. He studied pharmacy in both Iceland and Denmark and at one time was an assistant professor of pharmacology at the University of Iceland. After owning and managing several pharmacies in Iceland, he retired in 1999. It was then that Johannes started studying and photographing insects, which have always been a passion of his. He is still working at this pursuit.

Table of Contents

Introduction 3
1. What Are Insects? 4
2. How to Use This Book 6
3. Class Insecta 8
4. Adaptations and Cold-hardiness 13
5. Importance of Arctic Insects 15

CLASS INSECTA – INSECTS
   Orders   Ephemeroptera – Mayflies 20
                  Plecoptera – Stoneflies 22
                  Blattodea – Cockroaches 24
                  Phthiraptera – Lice 26
                  Hemiptera – True Bugs and Homopteran Insects 30
                         Aphididae – Aphids 34
                 Coleoptera – Beetles 36
                         Carabidae – Ground Beetles 38
                         Dytiscidae – Predaceous Diving Beetles 40
                         Hydrophilidae – Water Scavenger Beetles 44
                         Silphidae – Carrion Beetles 46
                         Staphylinidae – Rove Beetles 48
                         Elateridae – Click Beetles 50
                         Coccinellidae – Ladybird Beetles 52
                         Curculionidae – Weevils 54
                  Diptera – Flies 56
                         Trichoceridae – Winter Crane Flies 58
                         Tipulidae – Crane Flies 60
                         Culicidae – Mosquitoes 62
                         Simuliidae – Black Flies 66
                         Ceratopogonidae – No-see-ums 68
                         Chironimidae – Non-biting Midges 70
                         Sciaridae – Dark-winged Fungus Gnats 72
                         Cecidomyiidae – Gall Midges 74
                         Tabanidae – Horseflies 76

Empididae – Balloon Flies 78
  Dolichopodidae – Long-legged Flies 80
  Syrphidae – Hoverflies 82
  Scathophagidae – Dung Flies 84
  Anthomyiidae – Root-maggot Flies 86
  Muscidae – Houseflies 88
  Calliphoridae – Blowflies 90
  Oestridae – Warble and Botflies 92
  Tachnidae – Tachinid Flies 94
■ Lepidoptera – Butterflies and Moths 96
  Tortricidae – Leafroller Moths 100
  Pieridae – White and Sulphur Butterflies 102
  Lycaenidae – Gossamer-winged Butterflies 104
  Nymphalidae – Brush-footed Butterflies 106
  Pyralidae – Snout Moths 108
  Geometridae – Looper Moths 110
  Lymantridae – Tussock Moths 112
  Noctuidae – Owlet Moths 114
■ Trichoptera – Caddis Flies 116
■ Hymenoptera – Bees, Wasps, Ants, and Sawflies 118
  Tenthredinidae – Common Sawflies 122
  Braconidae – Braconid Wasps 124
  Ichneumonidae – Ichneumon Wasps 126
  Vespidae – Social Wasps 128
  Apidae – Bees 130

CLASS ARACHNIDA – ARACHNIDS
  ■ Araneae – True Spiders 138
    Lycosidae – Wolf Spiders 140
    Philodromidae – Crab Spiders 142
    Lyniphiidae – Sheet Web and Dwarf Spiders 144
Subclass Acari – Mites 146
    Some Interesting Mites 148

Glossary 150
Bibliography 155
Index 166

Introduction

Most of us are familiar with the large, spectacular wildlife that lives in Nunavut. Some of us have been lucky enough to see polar bears, caribou, muskoxen, and great flocks of seabirds in the wild. In this book, I am going to introduce you to a different world of animals: a world of complicated, diverse, fascinating, and tiny Arctic animals—insects and spiders!

Now, many readers probably think of insects and spiders with some level of disgust (such as when we think of maggots), irritation (such as when we think of mosquitoes), or fear (such as when we think of bees), but this reaction isn't entirely fair! For example, horror and adventure movies tend to make us foolishly scared of anything that creeps and crawls on the ground, in dark places, or near dead things. In reality, these small animals play essential roles in the Arctic ecosystem, especially in their relationships with plants, and are therefore fundamentally important to so many other animals in the North.

Insects and spiders are found from the southernmost to the northernmost parts of Nunavut. Unlike the mammals, birds, and freshwater fish in Nunavut, there are many species of Arctic insects and spiders that have not yet been named, and probably many that have not even been discovered. This means that there is a lot of room for new knowledge to be gained by keeping our eyes on the ground and learning about these tiny animals.

One important thing to remember is that insects and spiders are very different from one another—they just appear superficially similar. They are both part of the animal phylum Arthropoda (a huge group of segmented animals without backbones), which is the most diverse group of animals on the planet.

So get ready to learn about the most diverse and abundant group of animals in Nunavut! Keep an open mind as you read, and enjoy the tremendous range of adaptations and life histories that these tiny creatures exhibit in the cold climate of Nunavut.

Part One

What Are Insects?

Insects are the most numerous animals on the planet. More species of insects can be found in the Arctic than any other type of animal. They successfully live and reproduce in the harshest environments. Because of their size, however, most insects go virtually unnoticed—there are likely many more species yet to be discovered.

Insects belong to a group of animals called invertebrates, which means that they do not have backbones. Worms, snails, crabs, and spiders are also invertebrates. Insects are classified taxonomically in the Arthropoda phylum (all organisms are grouped into phyla based on their body characteristics). All arthropods, including spiders, insects, millipedes, centipedes, and crustaceans, have bodies with several segments and many pairs of jointed legs.

The phylum Arthropoda is divided into classes, which are smaller groups that have even more things in common. Insects make up one of these classes: class Insecta. Spiders belong to the class Arachnida, a separate class within the phylum Arthropoda. Although some spiders are included in this book, they will be discussed in a separate section.

Classes are further divided into orders, families, genera, and species. Each grouping becomes more specific until you end up with only one species. The language used to classify all living things is Latin. Specific insect names, just like plants or other animals, consist of two Latin terms: the genus name and the species name. For instance, *Homo sapiens* is the scientific name for people—*Homo* refers to the genus in which we are grouped, and *sapiens* refers to our exact species within that genus. Related organisms can have the same generic name, but each one has its own species name.

The following is a list of insects from this book and how they are organized in the grand scheme of the animal kingdom.

Kingdom Animalia

    Phylum Arthropoda
        Class Insecta
            Order Ephemeroptera
            Order Plecoptera
            Order Blattodea
            Order Phthiraptera
            Order Hemiptera
                Family Aphididae
            Order Coleoptera
                Family Carabidae

    Family Dytiscidae
    Family Hydrophilidae
    Family Silphidae
    Family Staphylinidae
    Family Elateridae
    Family Coccinellidae
    Family Curculionidae
 Order Diptera
  Suborder Nematocera
    Family Trichoceridae
    Family Tipulidae
    Family Culicidae
    Family Simuliidae
    Family Ceratopogonidae
    Family Chironimidae
    Family Sciaridae
    Family Cecidomyiidae
  Suborder Brachycera
    Family Tabanidae
    Family Empididae
    Family Dolichopodidae
    Family Syrphidae
    Family Scathophagidae
    Family Anthomyiidae
    Family Muscidae
    Family Calliphoridae
    Family Oestridae
    Family Tachinidae
 Order Siphonaptera
 Order Lepidoptera
    Family Tortricidae
    Family Pieridae
    Family Lycaenidae
    Family Nymphalidae
    Family Pyralidae
    Family Geometridae
    Family Lymantridae
    Family Noctuidae
 Order Trichoptera
 Order Hymenoptera

    Suborder Symphyta
      Family Tenthredinidae
    Suborder Apocrita
      Family Braconidae
      Family Ichneumonidae
      Family Vespidae
      Family Apidae
  Class Arachnida
    Order Araneae
      Family Lyniphiidae
      Family Lycosidae
      Family Philodromidae
  Subclass Acari

Latin is no longer spoken, so it does not change or evolve. Using an unchanging, globally shared language to name organisms allows scientists (and others) to be sure that they are talking about the same organisms. Common names for insects vary from language to language and from region to region in the same language; these names are not exact, and two different insects might have the same common name. An organism's scientific name in Latin is always italicized or underlined when written, and the genus name always begins with a capital letter. For example, *Aedes nigripes* is the Latin name for one of the mosquitoes that lives in Nunavut.

Part Two
How to Use This Book

Because there are so many different species of insects in the Arctic, this book will deal with insect orders and families rather than smaller classifications of species. Unless you are a real expert, it is hard to identify insects beyond the family to which they belong. Identifying an insect on the species level usually requires a microscope, as well as literature on, and expertise in, identifying organisms.

The orders in this book include the most common insects that are seen in the Arctic. These orders are broken down to include the most common families.

Each order or family account includes the following: English and Inuktitut names (where possible); a description of the adult and larva or nymph, the life cycle, the food and feeding, the habitat, and the range; a "Did You Know?" section of interesting facts; and traditional knowledge found during the research for this book. You will also notice a black drawing in the corner of each right-hand page. This drawing will help you to identify an insect by its general shape. Words marked in **bold** are included, with additional information, in the glossary at the end of the book.

For the purposes of this book, trying to define the exact locations where the insects were found did not seem to be realistic. Instead, the insects are simply listed as being found in the High or Low Arctic. The following map will help you to understand what the author means by the High and Low Arctic of Nunavut.

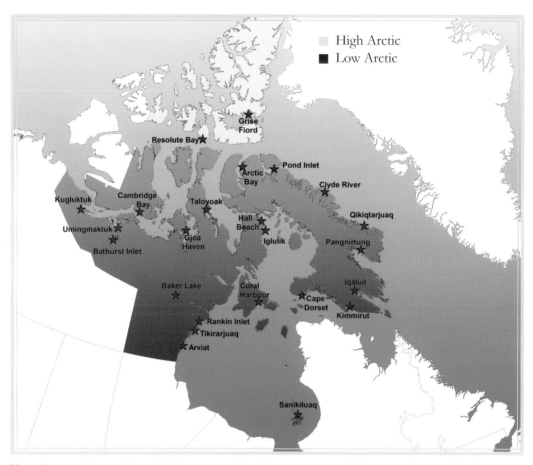

Figure 1

Part Three

A General Account of the Insect Class

Class: Insecta – Insects

What makes an insect an insect? Adult insects are distinguished from other animals in the phylum Arthropoda by their three body segments, six legs, one or more pair(s) of wings, and two antennae.

Description

Adults:
To become more familiar with insects, let's take a closer look at their body parts. The three distinct body sections of an insect are the **head**, the **thorax**, and the **abdomen**. All sections are covered by a hardened body wall called the **exoskeleton**.

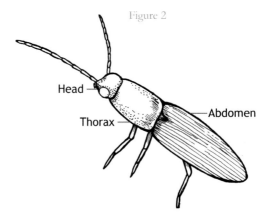

Figure 2

Head:
The head has two antennae, one to several pairs of eyes (one larger, prominent pair of **compound eyes** and perhaps one or more pairs of small, inconspicuous simple eyes), and varying mouthparts. The antennae are segmented and are usually found between or in front of the compound eyes. Antennae can look quite different depending on the insect species. They are used mostly for smell and touch, but in some cases, antennae can also be used for hearing.

Insects can have two types of eyes: compound eyes and **ocelli** (simple eyes). Compound eyes are the large pair that often cover most of the head. These eyes are composed of many facets (sometimes hundreds). The compound eyes are used to see things the same way that human eyes let us see them, although insect eyes do not have as good a resolution (ability to see fine details) as our eyes. Insect eyes are nonetheless capable of seeing ultraviolet light, and many flower patterns are only visible in ultraviolet. The ocelli are typically found in triangular trios on top of the heads of many adult insects. These simple eyes are very sensitive to light in general, and particularly to shifts in light. They help insects stay steady during flight and help regulate behaviour via changes in day length.

Mouthparts vary depending on the feeding habits of the insect. They are generally made up of an upper lip (**labrum**), a lower lip (**labium**), two jaws (**mandibles**), and two small, jaw-like appendages (**maxillae**). Mouthparts can be used either for biting or for sucking. Insects that bite their food do so by working their mandibles from side to side. Those that suck up their food have modified systems with beaks or long tongues. The mouthparts are very specialized. For example, mosquitoes use their mouthparts like prescise syringes, complete with injections of anti-coagulants and anaesthetics. Itching is our reaction to these injections, though usually the females are long gone with their blood meals by the time we feel the itch.

Biting mouthparts

Sucking mouthparts

Thorax:
The thorax, or mid-section, of an insect is itself divided into three sections. Each section has one pair of legs. The middle and last sections also each have one pair of wings (except in flies, which only have one pair of wings) and the muscles to accompany these. Exceptionally, some insects are wingless. On the sides of the thorax, there are two slits. These are the **spiracles**, the external respiratory openings.

Insect legs typically have six segments, as illustrated (Figure 4).

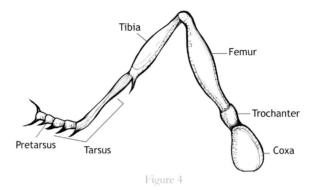

Figure 4

The wings usually have eight basic veins, the areas between which are called cells. Insects generally have two pairs of wings: the front wings and the hindwings. The front wings are attached to the middle thorax section, while the hindwings are attached to the last thorax segment. Typical wings are **membranous**, which means that they are thin, flexible, and almost transparent. There are adaptations in some of the orders. For example, beetles have a hardened pair of front wings called **elytra**. The hindwings of flies are very modified and no longer function as wings, but rather as stabilizers when they fly. These are called **halteres**.

Abdomen:
The typical insect abdomen, or third section of the insect's body, is divided into eleven segments. The last segment is often adapted to carry out a particular function. Thus, only ten segments are really visible as segments. In many insects, segments eight and nine form the reproductive structures.

Exoskeleton:
One of the important features of an insect, which allows it to be so successful, is its strong, durable, waterproof exoskeleton. The exoskeleton is a hardened body wall that protects the interior workings of the insect, provides support for the attached muscles, keeps the insect's softer parts from desiccating, and keeps the insect dry. Unlike mammals, birds, and reptiles, all arthropods (the phylum that includes insects) have their skeletons on the outsides of their bodies. Of course, one problem with having the skeleton on the outside is that this limits growth. Therefore, at times an insect's external skeleton is replaced by a newer, larger one.

The exoskeleton is comprised mostly of cuticle. The cuticle can take on many different forms. It makes up the hard, rigid outer covering of beetles, but it can also be more flexible and thin, as on caterpillars.

Breathing:
Insects have tubes that transmit air to all parts of their bodies. The tubes, known as tracheae, are open to the outside through spiracles on the sides of the body. The spiracles open and close through the use of small valves. Smaller tubes branch off the main tracheae.

Larvae and nymphs:
Immature insects are known as larvae or nymphs (depending on the insect group). Both larvae and nymphs hatch from insect eggs. Larvae undergo complete transformations in order to become adult insects, while nymphs do not. Larvae generally do not resemble the adults they will become, while nymphs look very much like the adults, only smaller.

Life Cycle

The insect life cycle includes a fascinating and unique feature called "metamorphosis." In this process, young insects hatch from eggs, and may look like small versions of the adults (**incomplete metamorphosis**) or like completely different types of creatures from the adults (**complete metamorphosis**). This process means that young insects may live in environments dissimilar to the ones they will live in when they are adults, and may eat different foods (unlike what you might expect of caribou, geese, or people). For example, young mosquitoes hatch in water and spend their early lives in ponds, feeding on fine organic materials, algae, and protozoans. Once they have pupated and changed to adults, they leave the water and fly over land to seek out blood meals.

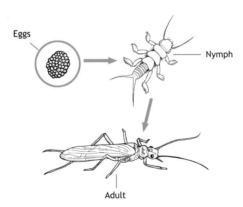

In the case of incomplete metamorphosis, the nymphs pop out of the eggs. Nymphs grow larger, and because they are encased in hard exoskeletons, they must **moult**, which means that they must shed their skins. After each skin change, they are closer to becoming sexually mature adults. Each stage of growth between the shedding of skins is called an **instar**. As they are growing, insects are referred to as first instar, second instar, and so on. The number of instars varies between orders and even families of insects. Once the nymph is at the last instar and final moult, the adult stage is reached (generally with wings).

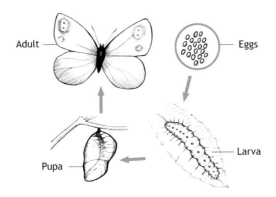

During complete metamorphosis, the larvae emerge from the eggs. Like the nymphs, larvae go through many instars, but they also experience one extra step in the development to adulthood. After the appropriate number of instars, the larvae enter a period of rest called the **pupal stage**. It is during this resting stage that the larvae transform into adult versions of the insect. When the adults emerge from the pupae, they will have wings and be sexually mature, allowing the cycle to begin again.

Food and Feeding

Insects consume a huge variety of foods and exhibit many feeding strategies. Some hunt for food, including other insects and their young, while some only eat plants. Some insects feed on dead materials, while others eat only live plants or animals. Some insects even live and feed off other animals or insects as **parasites**. Their feeding mechanisms and adaptations are aligned with the foods they eat.

Habitat

In Nunavut, there are limited habitats available to insects, especially compared to those available in southern areas (where there are more trees and marshes that insects can live in, for example). In Nunavut, insects can live on the tundra where there are plants, or in small pools, ponds, or lakes. A few larvae live in running water, as well. Vast rocky areas with little vegetation or water tend to support few insects.

Range

Insects are found all over Nunavut, in both the High and the Low Arctic. But the farther north you go, the fewer species you can find. The map (Figure 1) illustrates the boundaries of what is considered the High and the Low Arctic in the context of this book.

Traditional Knowledge

During the research for this book, elders in several communities in Nunavut were interviewed about insects. From these interviews, knowledge on certain insects was shared and has been included at the end of each account.

Part Four

Adaptations and Cold-hardiness

Most insects are cold-blooded. They have no internal heating mechanisms and cannot generate heat. To move around or fly, they need to be warmed up by the air temperature. Insects, like plants, need ways to cope with the short, cool summers, the generally unpredictable weather, and the long, cold winters of Nunavut.

Insects in the Arctic have developed cold-hardiness, which involves the use of many different adaptations. Most Arctic insects are freeze-tolerant, meaning that they can survive the winter while frozen. Much of the water in their bodies is converted to extracellular ice (ice outside of the cells). Insects continue to use some water in this frozen state, and 10–30% of their body water remains unfrozen. Some insects can survive temperatures of -60°C in this frozen state. Insects use glycerol (antifreeze), as well as some other chemicals, to protect their tissues during freezing and thawing.

One of the other common strategies that insects employ is freezing before the temperature of the air reaches the freezing point. This allows the insect to freeze more slowly than if the freezing were only controlled by outside temperatures.

Some insects **supercool** instead of freezing. The water in their bodies becomes very, very cold, but does not freeze. They also can survive unfrozen at -60°C, through the use of cryo-protectants (substances that prevent the freezing of tissues and that can prevent damage to cells during freezing).

Insects' resistance to dessication is another important factor in their survival against the cold. Because cold air does not contain much moisture, it can easily dry out insects, particularly small ones, so insects must protect the moisture in their bodies.

Freeze tolerance and resistance to dessication are vital physiological and biochemical adaptations found in Arctic insects, but there are other adaptations that are equally important in ensuring that insects can successfully **overwinter**.

Just like plants, insects go through the winter in sheltered locations, which increases their chances of survival. Such shelters can be under rocks, on the least windy sides of cliffs, or even under insulating layers of snow. Insects that are buried under snow are less likely to suffer from dessication caused by strong winds and wind-blown ice crystals or sand.

Similarly, **diapause** (suspended development particularly during unfavourable conditions) occurs in many Arctic insects before the terrible weather has a chance to harm them.

In addition to the adaptations needed for Arctic insects to survive the freezing, harsh winters, the cool, short Arctic summers require certain adaptations. Despite being cold-blooded, some insects can fly and develop at near-freezing temperatures. Some have higher metabolic rates, and, of course, many species have no choice but to develop rapidly. That being said, some species have also evolved to develop into adults over several seasons instead of just one.

Many Arctic biting flies do not need blood meals to produce their eggs. In fact, some of these flies no longer have mouthparts. They make their eggs through the use of food that they store in their bodies when they are larvae. This decreases the number of eggs they can produce, but it means that they do not need to locate warm-blooded creatures in the vast expanse of the Arctic.

You may have noticed that Arctic insects are both darker and hairier than southern ones. The dark colour absorbs heat from the sun and their hairs trap that heat. Many Arctic plants have evolved to have these adaptations. Another way that insects keep warm in the summer is by spending time in flowers. Some flowers follow the sun as it crosses the sky, and it is very common to see insects in the centre of these flowers, basking in the warmth.

Given that there is twenty-four–hour daylight in the summer across most of Nunavut, insects take advantage by being active whenever it is warm enough and there is little wind. Patterns of night and day belong only to their southern cousins.

Because summers are short, insects emerge as early in the spring as possible to complete their life cycles. Some stages of development can be stretched to last for more than one season if conditions are not right.

In other species, the need for a mate is no longer necessary. Females can produce eggs on their own without fertilization, leaving nothing up to chance. This process is called **parthenogenesis**.

As you can see, many interesting adaptations have been necessary for insects to survive harsh Arctic conditions.

Part Four
Importance of Arctic Insects

When you think of insects in the Arctic, the first ones that probably come to mind are mosquitoes and blackflies. While there are lots of those in Nunavut, there are many other insects that play critical roles in Arctic ecosystems. Two of the most important functions of insects are the pollination of plants and the breakdown of detritus. Many species of Arctic plants rely on insects (especially flies) to transfer pollen from flower to flower, allowing for plant reproduction and the creation of fruit such as berries. Also, insects do the essential duty of helping to eat and break down dead organic material, such as dead leaves, dung, and the carcasses of animals. This service helps return nutrients to the soil so that plants and other life can grow. Without insects, dead animals, along with dung, decaying plants, and even seaweed, would be piling up on our landscape! Yet, while insects are vital to life in the Arctic, much remains to be learned and appreciated about them.

Class Insecta
Insects

Orders
- Ephemeroptera – Mayflies 20
- Plecoptera – Stoneflies 22
- Blattodea – Cockroaches 24
- Phthiraptera – Lice 26
- Hemiptera – True Bugs and Homopteran Insects 30
  - Aphididae – Aphids 34
- Coleoptera – Beetles 36
  - Carabidae – Ground Beetles 38
  - Dytiscidae – Predaceous Diving Beetles 40
  - Hydrophilidae – Water Scavenger Beetles 44
  - Silphidae – Carrion Beetles 46
  - Staphylinidae – Rove Beetles 48
  - Elateridae – Click Beetles 50
  - Coccinellidae – Ladybird Beetles 52
  - Curculionidae – Weevils 54
- Diptera – Flies 56
  - Trichoceridae – Winter Crane Flies 58
  - Tipulidae – Crane Flies 60
  - Culicidae – Mosquitoes 62
  - Simuliidae – Blackflies 66
  - Ceratopogonidae – No-see-ums 68
  - Chironimidae – Non-biting Midges 70
  - Sciaridae – Dark-winged Fungus Gnats 72
  - Cecidomyiidae – Gall Midges 74
  - Tabanidae – Horseflies 76
  - Empididae – Balloon Flies 78
  - Dolichopodidae – Long-legged Flies 80
  - Syrphidae – Hoverflies 82

- Scathophagidae – Dung Flies 84
- Anthomyiidae – Root-maggot Flies 86
- Muscidae – Houseflies 88
- Calliphoridae – Blowflies 90
- Oestridae – Warble and Botflies 92
- Tachnidae – Tachinid Flies 94

■ Lepidoptera – Butterflies and Moths 96
- Tortricidae – Leafroller Moths 100
- Pieridae – White and Sulphur Butterflies 102
- Lycaenidae – Gossamer-winged Butterflies 104
- Nymphalidae – Brush-footed Butterflies.......................... 106
- Pyralidae – Snout Moths 108
- Geometridae – Looper Moths.. 110
- Lymantridae – Tussock Moths 112
- Noctuidae – Owlet Moths 114

■ Trichoptera – Caddis Flies 116
■ Hymenoptera – Bees, Wasps, Ants, and Sawflies 118
- Tenthredinidae – Common Sawflies 122
- Braconidae – Braconid Wasps 124
- Ichneumonidae – Ichneumon Wasps................................ 126
- Vespidae – Social Wasps... 128
- Apidae – Bees... 130

Mayflies

Order: Ephemeroptera – Mayflies

The insects in this order are commonly known as mayflies in English. No Inuktitut names were found for this order during the research for this book. If you know of any, please contact the Nunavut Teaching and Learning Centre.

The name Ephemeroptera comes from the Greek roots *ephemeros* ("short life") and *pteron* ("wing"). Once mayflies become adults and their wings dry, they fly up into the air to mate, and then the males die. The females only live long enough to deposit their eggs. These insects have been around for the last 350 million years.

Description

Adults:
Mayflies are yellow or brownish in colour. They have soft bodies, large heads, and very noticeable eyes that cover most of their heads. They have long, thin legs. Their mouthparts are not fully functional because they do not eat as adults. They have two pairs of wings: the front wings are large and triangular with many veins, and the hindwings are much smaller and rounder. Mayflies have small, stiff, barely noticeable antennae. As well, they have two or three **cerci** (feeler-like appendages at the hind end of the abdomen) that are about twice the length of their **abdomens**.

Nymphs:
Nymphs—mayflies in their immature stages—are **aquatic**, meaning that they live in water. The nymphs have biting mouthparts, unlike the adults whose mouthparts are not fully functional because mayflies do not eat during the short adult period of their lives. For breathing, the nymphs have **gills** (respiratory organs for breathing under water) on their abdomens. They also have three cerci.

Life Cycle

Incomplete Metamorphosis:
Once they emerge from the nymph stage, the males fly up and down in swarms. When females join the swarms, they are grabbed by the males, and mating proceeds while in flight. Eggs are laid within an hour, either on aquatic plants or in the water. The males die shortly after mating, while the females die shortly after depositing the eggs. Most of the adults live for less than a day. Once the eggs hatch, the nymphs go through ten to fifty **moults**, which

can take anywhere from several weeks to one year. In the last aquatic stage, the insects leave the water, moult, and emerge with smoky-looking wings. This stage is called the **subimago** and is the final stage before adulthood. The subimagos quickly moult one more time, shedding their outer skins to become adults with clear wings. The subimago stage of mayflies is unique: no other insects shed their skin once they already have wings.

Food and Feeding

Mayfly nymphs feed on plants or small aquatic animals. The adults do not feed at all.

Habitat

Most nymphs live in actively moving freshwater, although there are a few species that can survive in brackish water (water with some salt content). Some nymphs swim, while others crawl on the muddy bottom or dig burrows.

Photo by Richard Bartz

Range

Mayflies are found all over the world except in Antarctica and the High Arctic. You might see these insects around Nunavut if you live in a Low Arctic community.

Did You Know?

Mayfly nymphs can only live in unpolluted water. Therefore, they are good ecological indicators for scientists. If mayflies continue to thrive in a river or lake, then scientists can be fairly sure that the water in that lake or river is in good condition.

Stoneflies

Order: Plecoptera – Stoneflies

The insects in this order are commonly known as stoneflies in English. They are called stoneflies because they are often found resting or crawling on rocks near water. Plecoptera is derived from Greek and means "folded wings." No Inuktitut names were found for this order during the research for this book. If you know of any names for species in this order, please contact the Nunavut Teaching and Learning Centre.

Description

Adults:
The long, flat adults are soft-bodied, and have four **membranous** wings. The wings in the front are long and narrow, while the hindwings are shorter, with wider, rounder parts at each end. The wings sit flat on the **abdomen** when the insect is at rest. Stoneflies can be anywhere from three to fifty millimetres in length, and have long, threadlike antennae. The adults also have two **cerci** that are made up of many segments. Stoneflies have chewing mouthparts, but they are not well-developed.

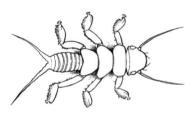

Nymphs:
Of stoneflies, as of mayflies, the immature forms are called **nymphs**. They are **aquatic**, and have elongated bodies, lengthy antennae, and slender cerci. Stonefly nymphs have threadlike **gills** on their **thoraces**—at the base of the legs, on the neck, or on the abdomen.

Life Cycle

Incomplete Metamorphosis:
Mating occurs during the day. To attract a mate, the male stonefly sometimes beats his abdomen on the ground. A female who does not yet have a partner will respond by beating her abdomen on the ground. Once they locate each other, they mate. Afterwards, while in flight, the female stonefly dips her abdomen in the water and lays a mass of eggs. The eggs become sticky, sink to the bottom, and adhere

Photo by Bill Stark

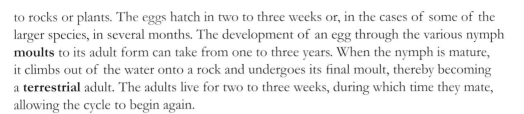

to rocks or plants. The eggs hatch in two to three weeks or, in the cases of some of the larger species, in several months. The development of an egg through the various nymph **moults** to its adult form can take from one to three years. When the nymph is mature, it climbs out of the water onto a rock and undergoes its final moult, thereby becoming a **terrestrial** adult. The adults live for two to three weeks, during which time they mate, allowing the cycle to begin again.

Food and Feeding

Most adults have poorly developed **mandibles** and do not eat at all. In some species, however, the adults eat blue-green algae or plants. Depending on the species, the nymphs can either be **predators**, hunting and eating other animals, or eaters of **detritus**. The predators actively search and pursue other aquatic **arthropods**, such as insect nymphs or small crustaceans. Some swallow the prey whole, while others bite off pieces and swallow those pieces whole. Detritus-eaters feed on decaying plant and animal material that has become caught on the stream or lake bed. They shred the material into smaller pieces with their mandibles.

Habitat

The adults are never found far from water since they are only alive for a short time and must return to water to lay their eggs. The nymphs need well-aerated water (water with lots of air) to thrive.

Range

As far as we know, these insects occur everywhere in the world except Antarctica and the High Arctic, so they are found in and around the Low Arctic lakes and rivers of Nunavut. Elders from Baker Lake have reported seeing insects from this order in their area.

Did You Know?

Even though the insects in this order have wings, most of them are poor fliers and will run rather than fly when disturbed.

Photo by Bill Stark

Order: *Plecoptera*

Cockroaches

Order: Blattodea – Cockroaches

Members of the Blattodea order are commonly known as cockroaches. No Inuktitut names for this order were found during the research for this book. If you know of any, please contact the Nunavut Teaching and Learning Centre.

Cockroaches are adventitious, which means that they can survive in places where they are not usually found. Because the climate in Nunavut is too cold for these insects, they do not naturally occur there, but some of them were brought there and have managed to survive.

Description

Adults:
Cockroaches are typically between three and eighty millimetres long. Each has a flat, oval head and a mouth built for chewing, biting, and licking. The head is partially hidden because, like the **thorax**, it 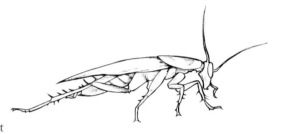 is covered by a **pronotum** (hardened cover). The cockroach has long, whisker-like antennae that are often swept back along the sides of its head and thorax, and two **compound eyes**. Cockroaches come in a variety of colours, ranging from black to brown and sometimes even green and red. They have hard, thick front wings and flexible, thin hindwings. Their wings overlap one another, unlike beetles, whose wings meet exactly in the middle of their

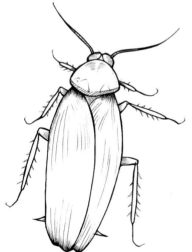 **abdomens**. Cockroaches' short legs are adapted for running quickly, so, although they have wings, they rarely fly. They have short, segmented tails or **cerci**. The cerci behave like antennae at the insect's back end, helping make it aware of its surroundings.

Nymphs:
Young cockroaches look much like the adults, but they are lighter in colour, have shorter antennae, and have no wings. They have **wingpads** on their thoraces, which metamorphose into wings as the nymphs becomes adults. Young cockroaches are softer than the adults, but they harden as they are exposed to air.

Life Cycle

Incomplete Metamorphosis:
The female cockroach lays twelve to twenty-five eggs in a special capsule called an **ootheca**. Once she lays the eggs in the capsule, she either glues the capsule to the ground in a safe location or attaches it to her hind end. In a few cases, the ootheca stays inside the female's body until the young are ready to hatch. The eggs can take anywhere from a few weeks to a few months before they are ready to hatch. To get out of the ootheca, the cockroach nymphs inflate themselves by swallowing air. The pressure causes the ootheca to split open and the little cockroaches to hatch. They go through two to twelve **moults** before they become adults. An individual cockroach may live for several years.

Food and Feeding

Cockroaches will eat almost anything. This type of feeding behaviour is described as being **omnivorous**. They are also **scavengers**, meaning that they do not usually hunt for their food, but instead eat dead things that are lying around. They come out at night to look for food, and spend the rest of their time hiding in small crevices. Their flat bodies allow them to hide almost anywhere.

Habitat

Most of the cockroaches that we are familiar with are pests. Their habitats are houses and other buildings. Other cockroaches live on the ground, where they can hide under rocks and logs or in caves during the day and come out at night to feed.

Range

Cockroaches live mostly in the tropics, where it is very warm. The more northern species are the pests that have taken advantage of heated houses and other buildings. In Nunavut, cockroaches have only been found at the military base in Alert on Ellesmere Island, where it is assumed that they arrived with some military cargo.

Did You Know?

In the insect world, the capsules in which eggs are laid, known as oothecae (plural of ootheca), are only made by cockroaches and praying mantids. This is one reason why these two different insects are considered related. There are fossils of cockroaches in rocks from 305 million years ago.

Lice

Order: Phthiraptera – Lice

The insects in this order are commonly known as lice (singular: louse) in English. Lice are known as *kumaq* and *iqqiq* in Inuktitut.

Historically, lice were divided into two orders: Mallophaga (biting lice) and Anoplura (sucking lice). They are similar enough, however, that many entomologists now group them together. We will discuss both groups of lice here.

Description

Adults:
Lice are small, wingless ecto-parasites. Ecto-parasites are **parasites** that live on the outside of the bodies of their **hosts**, as opposed to on the inside. The host species receives no benefit from this relationship and usually experiences some harm from it.

Lice are generally less than ten millimetres long. Their bodies and heads are flat so they can cling close to their hosts' skin. They are mostly colourless and have very small eyes or no eyes at all. The last segments of their legs are hooked, which helps them cling to their hosts. Lice have no wings and therefore must walk on their hosts—they cannot fly or jump. Lice can have either chewing or sucking mouthparts. They have short antennae and are covered in sensory hairs, which help them to feel their way around since they have poor to no eyesight.

Biting lice have larger heads than sucking lice, although sucking lice are larger overall.

Nymphs:
Lice **nymphs** look like adult lice except that they are smaller and have curled legs that have been modified for hanging onto their hosts. The nymphs do not develop leg hooks until the adult stage.

Life Cycle

Incomplete Metamorphosis:
Female lice lay eggs every day on their hosts until they have laid between one and three hundred eggs. In the egg stage, the insect is called a nit. The female glues the nits to the hairs or feathers of the host with a cement-like substance that she produces. The top of each nit has a cap. To get out of the egg, the louse swallows air, which forces the cap off due to the increased air pressure inside. Sometimes the pressure is so great that the louse gets shot out of the egg! Lice do not live very long—usually from two to eight weeks. Lice are unable to survive if they are off their host for more than twenty-four to thirty-six hours.

Food and Feeding

Sucking lice feed on the blood of the mammal or bird that they are living on. Blood alone is not a well-balanced diet, so these lice have bacteria that live in their intestinal systems and provide them with additional nutrients. Most biting lice feed on fur and feathers, although some species take advantage and feed on the blood produced when their host scratches to try to get rid of them.

Habitat

The habitats of lice are their host species (the birds and mammals on which they live). Some lice live on one part of their host's body their whole lives. Biting lice mostly live on birds, although there are a few that prefer to live on mammalian hosts. Sucking lice only live on mammals and there are several species of sucking lice that make their homes on human beings. *Pediculus humanus capitis*, also known as head lice, live on the human scalp. *Pediculus humanus corporis*, also known as body lice, live on the human body and lay their eggs in our clothing. *Phthirus pubis*, also known as pubic lice or "crabs," are squatter and broader than the other two and they move around less.

Range

Lice are found wherever there is an appropriate host to live on. They live in the High and Low Arctic.

Did You Know?

Pediculosis occurs when large numbers of lice live on any part of the human body. The symptoms are skin irritations, possible allergic reactions, and a general feeling of physical discomfort like that at the beginning of an illness.

Order: Phthiraptera

Traditional Knowledge

From Iqaluit: Sylvia Inuaraq interviewing Pauline Erkidjuk, July 2006

The colour of head lice is black. The color of lice on the body is white. The larvae of head lice stick to the hair. When Inuit [used to] pick lice from the head, they usually ate the lice by squishing them with their teeth or by using their nails. When Inuit [used to] use caribou skin clothing, the lice on the caribou skin clothing were white.

From Taloyoak: Ellen Ittunga interviewing Judas Ikadliyuk, July 2006

When I was a child, people and also caribou bedding had lice. We do not see lice anymore because today we have a healthy lifestyle. We have doctors and nurses to control this. I have seen head lice grow very big. They are black compared to bedding lice, which are whitish or lighter in colour. The head lice larvae are called itqiit. *They are considered head lice waste, which we call* anaq. *They are found in the hollow part of the hair.*

From Taloyoak: Ellen Ittunga interviewing Lucy Ikadliyuk, July 2006

We used to find caribou lice on our bedding; we used to have to shake our bedding daily outside our iglu. Head lice are dark in colour, while animal lice are light in colour. They tend to be whitish. Today, I do not see any more kumak *like we did many years ago.*

From Iqaluit: Sylvia Inuaraq interviewing Celestine Erkidjuk, July 2006

Back then, when a person had lice and was unhealthy, people could tell if the person was going to live or not by the direction of movement of the lice.

Order: *Phthiraptera*

True Bugs and Homopteran Insects

Order: Hemiptera – True bugs and homopteran insects

Taxonomists generally agree that order Hemiptera includes four suborders: the suborder Heteroptera (true bugs) and the three homopteran suborders: Sternorrhyncha, Clypeorrhyncha, and Auchenorrhyncha.

Description

Adults:
Insects of the order Hemiptera range from one millimetre to eleven centimetres long. They come in various colours, most commonly browns and greens.

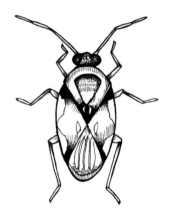

A true bug (an insect of the suborder Heteroptera) has sucking mouthparts on a beak at the front of its head. The beak is made up of a long **rostrum** (snout-like extension of the mouth), which gets thinner as it projects away from the head. Two pairs of **stylets** (needle-like structures) fit into a groove in the rostrum—the first pair is equipped with teeth to cut up the food, while the second pair is structured to suck the food in. When the rostrum is not in use, it is held back between the two front legs. When in use, it might be stabbed into a plant, another insect, or even a larger animal.

Insects of the other suborders—that is, the homopteran insects—have thin beaks pressed very close to the bottoms of their heads, making them look either like they come out from the backs of the heads or from between the front legs.

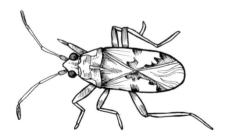

All the insects in this order have large **compound eyes**, and some also have a pair of **ocelli**.

The front wings of most true bugs are thickened from the base, where they attach to the body, to halfway down the wings. Then, from the middle to the tip, the wings are membranous. The hindwings, used for flying, are completely membranous and are shorter than the front wings. When the bugs are resting, both sets of wings are held flat over their bodies, with the tips of the front wings overlapping the hindwings. Most homopteran insects hold their

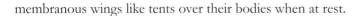

membranous wings like tents over their bodies when at rest.

Just to complicate matters, there are also quite a number of wingless insects in the Hemiptera order.

In true bugs, the third segment of the **thorax** is enlarged and triangular. It acts as a hard shield and is called the **scutellum**. This is characteristic only of true bugs, not of homopteran insects.

Some insects in this order are **aquatic**, and some are **terrestrial**. The aquatic species have short and hidden antennae, while the terrestrial species can have short or long antennae. The legs of the insects in this order are mostly used for walking and running, although some true bugs have modifications for jumping and swimming.

Nymphs:
The **nymphs** look like the adults, although their colouring may be different. Also, they have either no wings or just **wing buds**.

Photo by Henri Goulet

Life Cycle

Incomplete Metamorphosis:
Some species in the order Hemiptera have mating rituals, while others simply find each other and mate without ritual. The females can lay anywhere from a few to many eggs. The eggs are inserted into plant tissue, glued to plant surfaces, or laid on soil or stones. In some species, the parents guard the eggs. Many species of aphids also reproduce **parthenogenetically** during certain stages of the life cycle, giving birth to live young. The insects in this order go through five stages of metamorphosis, but they do not go through a pupal stage, so the metamorphosis remains incomplete.

Food and Feeding

Insects in this order feed mostly on plant juices, which they suck up through their rostra. Some, however, also feed on other insects or small **vertebrates**.

Habitat

Most hemipterans are terrestrial, generally living on plant leaves. Some hemipteran species are aquatic, while others are semi-aquatic, which means that they live on or near water, but not underwater.

Range

About one dozen species of plant bugs (Miridae family) occur in the Low Arctic of Nunavut. A couple of species of predatory shore bugs (Saldidae family) have been collected in the Low Arctic and one species in the High Arctic. About eight species of leafhoppers (Cicadellidae family) have been identified in the Low Arctic. At least one species of short-winged planthoppers (Dephacidae family) has been found in the Low Arctic. Aphids (Aphididae family) are by far the most common homopteran insects in the Low Arctic, with a few species also being identified in the High Arctic. There are a couple of scale insects and mealybugs that have been found both in the High and Low Arctic (Coccidae and Pseudococcidae families).

Did You Know?

The word "bug" is generally used to describe any insect, but the name was actually first used to describe insects in the order Hemiptera. As taxonomists continue their research, they have found that the order Hemiptera now contains three or four suborders, but only members of the suborder Heteroptera are still referred to as "true bugs."

Order: Hemiptera

Aphids

Order: Hemiptera, suborder Sternorrhyncha; Family: Aphididae – Aphids
These tiny insects are commonly called aphids in English. In Iqaluit, aphids are identified as *kumaru* and *qupirrualaat* in Inuktitut. In Clyde River, they are called *quaraq* and *kumak*, while in Baker Lake, an aphid is a *kumaruq*.

Aphids are **terrestrial**. They are plant **parasites**, which means that they live on plants but contribute nothing to them in return. Depending on the species, they can be found living on almost any part of the plant from the roots up.

Description

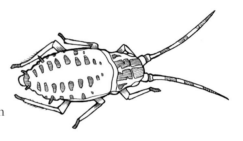

Aphids are small insects, generally ranging from one to five millimetres in length. They are soft-bodied and pear-shaped. They have small heads and six-segmented antennae. Aphids are usually wingless, although there are some winged forms. As with all other hemipterans, they have sucking mouthparts. Aphids are usually green, but they can also be red, black, brown, yellow, pink, and even transparent.

These small insects have several interesting features. Each has a **cauda** (tail-like structure) at the tip of its abdomen, which it uses to get rid of the excess sugar and water it takes in. The faeces that result from the excess sugar and water are called **honeydew**. Aphids have two small tubes on the hind ends of their abdomens called **cornicles**. These tubes produce a waxy substance used for defense, and also secrete alarm pheromones (chemicals produced to communicate with other members of the same species).

Life Cycle

Incomplete Metamorphosis:
Aphids **overwinter** mostly as eggs. In the spring, the eggs hatch into wingless female aphids. These wingless females give birth **parthenogenetically** to live, wingless females. After a series of such births, the wingless females give birth to winged females who may leave the colony to start a new colony elsewhere. It is thought that this cycle of births is shorter in the Arctic than in more southern areas. In the fall, these

Photo by Johannes F. Skaftason

winged females give birth to wingless males and egg-laying females. These last two mate. Soon after, the female lays eggs and the cycle starts again as the new eggs overwinter.

Food and Feeding

Aphids feed on the juices found in plant leaves, flowers, stems, and roots. The plant juices that aphids suck up have more sugar than protein. To get enough protein, aphids drink much more sugary juice than they need. The excess sugary fluid is excreted through their caudae. This sugary fluid (honeydew) becomes food for many other insects. For instance, bees and ants have been known to feed on this honeydew.

Photo by Johannes F. Skaftason

Habitat

Aphids live on plants, such as saxifrages, willows, blueberries, chickweed, mountain avens, and fireweed.

Range

About twenty species of aphids have been reported across Nunavut. Only three species have been recorded in the High Arctic so far.

Did You Know?

Ants take feeding on honeydew a step further than most other insects do. They actually "milk" aphids like cows in order to obtain their honeydew. In return for the honeydew, the ants protect the aphids from predators.

Photo by Johannes F. Skaftason

Beetles

Order: Coleoptera – Beetles

The insects in this order are commonly known as beetles in English. The Inuktitut word *minguq* seems to designate any type of beetle, although some beetles also have specific names.

The order Coleoptera is not only the largest order in the insect world, but also the largest in the whole animal kingdom, describing some 350,000 species. There are 125 families of beetles within this order. Species from about eighteen families have been identified in the Arctic.

Description

Adults:
Beetles range from 0.25 millimetres to seventeen centimetres in length. The main feature that distinguishes a beetle from other insects is its **elytra**. When a beetle is at rest, the two halves of the elytra come together in a straight line along the beetle's back, covering its hindwings. The hindwings are **membranous** and are used for flying. When the beetles are in flight, the hard elytra are held out at the sides of their bodies, while the hindwings help do the work of flying. Beetles have antennae with ten or eleven segments. They also have large **compound eyes** and chewing mouthparts with well-developed **mandibles**.

Larvae:
There are three types of beetle larvae: some of them are slender and crawly; others are grub-like and fleshy; and some, known as wireworms, are elongated and have tiny legs and hard **exoskeletons**. All beetle larvae are soft-bodied and have well-developed heads and chewing mouthparts. None of the larvae resemble the adults they will become.

Life Cycle

Complete Metamorphosis:
In most cases, beetles begin life as eggs and then become larvae. The larvae go through several **instars** followed by **moults**. This is followed by a period of rest as **pupae**, during which the beetles transform into adults and develop wings. Beetle pupae do not feed. Adult beetles emerge from the pupae, usually in the summer, and preparations then begin for the life cycle to be repeated.

Photo by Carolyn Mallory

Food and Feeding

Beetles are an incredibly diverse group of insects, and they feed on a wide variety of foods. Some beetles are **herbivores**, which means that they eat only plant material, while others eat only fungi. Some beetles are **predaceous**, hunting and eating other insects or small organisms, such as snails or worms. Still other beetles eat only dead matter—dead plants or animal carcasses.

Beetles have chewing mouthparts, so no matter their feeding preferences, they eat by biting and chewing the food into small bits.

Habitat

Once again, because of their great diversity, beetles are found almost everywhere. You might find them scurrying across the tundra, crawling under rocks, burrowing into the soil, living on plants, or swimming in small ponds. Some beetles are even **parasites**, living on host organisms and contributing nothing in return.

Range

Many species of beetles can be found in the Low Arctic, while only a few species of beetles have made it as far north as the High Arctic. Several species of diving beetle, one species of ground beetle, and a few species of rove beetles have been reported by entomologists in the High Arctic.

Did You Know?

Water beetles live and feed in ponds. But what happens if the pond dries up or if there is no more food? These beetles then fly to another puddle or small pond. Like all other beetles, water beetles have wings, even though we rarely think of them as being able to fly.

Photos by Henri Goulet

Order: Coleoptera

Ground Beetles

Carabidae

Order: Coleoptera – Beetles

Family: Carabidae – Ground Beetles

In Iqaluit, ground beetles are known as *qaurulliq* and *qalirualik*. In Clyde River, they are also known as *qalirualik*, while in Baker Lake, the name is *qalirulik*.

As is suggested by their English common name, ground beetle adults live mostly on the ground, where they are often found under rocks, particularly near water.

Description

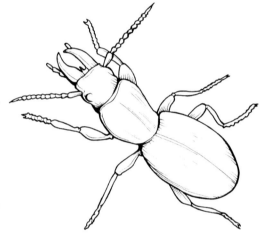

Adults:
Ground beetles can measure anywhere from three to sixty millimetres in length. Most of the insects in this group are shiny and black, but some can be brightly coloured. Ground beetles have narrow heads with large **compound eyes** and threadlike antennae, which are attached between the eyes and the **mandibles**. They have long, slender legs and their bodies are somewhat flat. Their **elytra** have lines of small punctures that are sometimes coloured and can appear grooved.

Larvae:
A great many of the carabid larvae have not yet been described. Some of the larvae that have been described are elongated, and have prominent antennae with three to four segments, as well as prominent mandibles. Their legs have six segments.

Life Cycle

Complete Metamorphosis:
The eggs are usually laid on objects above ground or in holes made in the soil. It generally takes the eggs five days to hatch. The larvae go through three **instars**, during which they live in burrows in the soil or among leaves and twigs on the ground. After three instars, the larvae **pupate**

Photo by Henri Goulet

underground. The adults emerge after five to ten days. These beetles can **overwinter** as adults, in which case they are ready to lay eggs in the spring and start the cycle again. When they overwinter as adults, they can be found in the soil or in protected areas. Or they may overwinter as eggs, in which case they hatch in the spring.

Development from an egg to an egg-laying adult takes about one year. In harsh conditions, the development can take longer.

Food and Feeding

Most of the beetles in this family are **nocturnal**. Both the adults and the larvae are **predaceous**. It is thought that in the Arctic, they mostly eat smaller insects that can be easily overpowered (for example, the larvae of other groups of insects). There is, however, one species from the High Arctic that is known to eat flowers. In general, they are **scavengers**.

Regardless of what they eat as adults, they use their sharp mouthparts to chew up their food. The larvae have different feeding mechanisms. They inject digestive juices into their food and then suck up the partially digested food in a liquid form. In this manner, they are somewhat similar to predaceous water beetles (Dytiscidae). This process is called **external digestion**.

Habitat

Ground beetles can be found in cracks in the soil, under rocks or plant debris, or even just running along the ground. Since they are active at night, they are most likely hiding under something during the day.

Range

There are eighty or more Arctic species of ground beetles. Most of the thirty-one species found in Nunavut so far have been found close to the treeline or in the southernmost tundra. Elders from Baker Lake, Clyde River, Iqaluit, and Kugluktuk have all identified ground beetles in their areas. Only one species is known to reach the High Arctic.

Did You Know?

If ground beetles are disturbed, they try to run quickly in order to escape, instead of flying. In fact, they seldom fly, particularly not in the Arctic.

Photo by Henri Goulet

Order: *Coleoptera*

Predaceous Diving Beetles

Dytiscidae

Order: Coleoptera – Beetles

Family: Dytiscidae – Predaceous Diving Beetles

The beetles in the Dytiscidae family are commonly known as predaceous diving beetles in English. They are **carnivorous** water beetles.

All water beetles are named *iqqamukisaat* in Igloolik and Clyde River, from the word *iqqaq*, which means "deep in the water." In Kugluktuk, they are simply referred to as *kingok*, or "bugs from the lake." In Baker Lake, elders identified these beetles as *pootoogooqsiut*, while in Iqaluit they are called *aqammukitaaq*, *ikarmikitaa*, or *imarmiutait*. In all of the communities, elders also simply called them *minguq*, the generic name for any type of beetle.

Description

Adults:
Predaceous diving beetles are smooth, elongated, and oval, and range from one to forty millimetres long. Like all other beetles, they have **elytra** covering their abdomens. They have flat hind legs with fringes of hair that act as swimming paddles. These beetles have threadlike antennae.

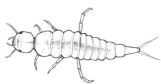

Larvae:
The larvae are often called water tigers. As with all beetles, the larvae look nothing like the adults. They are thin and tapered, with paired, tail-like appendages called **urogomphi**. The larvae have twelve segments—the head, three on the **thorax**, and eight on the abdomen—and long legs. They are equipped with long, crescent-like **mandibles**.

Life Cycle

Complete Metamorphosis:
Most dytiscids **overwinter** as adults and lay eggs in the spring, but some spend the winter as larvae and then change into adults in early summer. The most typical way for dytiscids to lay eggs is by making small cuts in plant tissue and laying their eggs inside the cuts. In this way, they use **aquatic** plants to supply oxygen to their eggs. Other methods include scattering the eggs on mud or in rubble by the edge of a body of water, or simply laying the eggs singly on leaves floating in the water. The eggs take a few days to a few weeks to hatch,

Photo by Johannes F. Skaftason

depending on weather conditions and water temperature.

Once the larvae come out of the eggs, they go through several **instars** followed by **moults**. Each time the larva moults, it grows a little bigger. The final instar crawls out of the water, digs a small hole in the ground, and becomes a **pupa**, inside which the wings and adult body develop. Once the adult form is ready, the dytiscid emerges from the pupa and flies or crawls back to the water to begin its adult life. Most dytiscids reproduce only once per year.

Food and Feeding

Both adults and larvae are predaceous, meaning that they hunt and kill prey, which they then eat. They feed primarily on small aquatic animals, such as the larvae of other species, tadpoles, molluscs (snails, for example), small frogs, and even small fish.

The adults feed on both living and dead or dying organisms by biting off small pieces and swallowing them a little bit at a time.

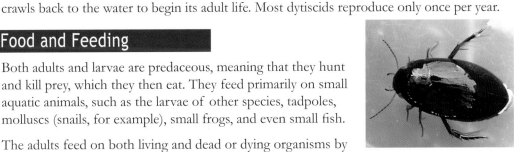

Photo by Johannes F. Skaftason

The larvae inject their prey with digestive enzymes using their long, hollow mandibles, and then they suck out the contents and discard the body. They are the vampires of the insect world!

Habitat

Dytiscid beetles live in quiet bodies of water, such as large puddles, ponds, lakes, and some slower sections of moving water.

The adults are terrific swimmers. They can be distinguished from similar-looking water beetles because they swim with both hind legs simultaneously, doing a frog-like kick. Other water beetles swim with their legs stroking alternately. All adult water beetles can stay underwater for long periods because they store air in chambers under the elytra. If you look closely, you will see them come up to the surface and, with their hind ends at the surface, replenish their oxygen supplies.

Some dytiscid species have larvae that swim, while others have larvae that cling to leaves or crawl along the bottom of the ponds in which they live. The larvae do not have the same capability as the adults do to store air, so they must come to the surface more often. Some larvae have **syphons** that they send up to the surface for air. The adults can fly quite well and will search for new habitats if their ponds dry up. They fly at night and are attracted to the reflectiveness of water.

Range

About twenty-six dytiscid species have been reported in the Arctic, with about half of these being found in the Eastern Arctic. Two species of dytiscids have been found in ponds in the High Arctic. Elders interviewed in the Baker Lake, Iqaluit, Clyde River, and Kugluktuk areas recognized this water beetle.

Did You Know?

To a certain extent, predaceous diving beetle larvae and other insect larvae are capable of replacing or regrowing body parts that have been lost. These parts, such as legs or antennae, can only be partly replaced by moulting, but apparently can be completely regrown in the pupal stage. It seems that the chance of completely regrowing a part is greater if the part is small and if the larva loses it while very young.

Traditional Knowledge

There is a little song that was collected by Vladimir Randa about *iqqamukisaaq* and it goes like this: *iqqamukisaaq itirmigut nuijarli!*

From Clyde River: Rebecca Hainnu interviewing Lydia Jaypoody, July 2006

> Lydia: *Once there was a person who was out walking. She was very thirsty. It was dark when she took a drink of water. She felt something but she thought it might be a piece of plant. Her stomach swelled up and one could hear the insect munching on her. In those days there were shamans and they would chant for the insect to change into a child in the woman. When the woman gave birth, it was born as a child but [with] long front teeth.*
>
> Rebecca: *Really.*
>
> Lydia: *That is how it was told. People are discouraged from consuming them through drinking water. There are other ones occasionally found in water tanks. They are known as* iqqamukisaaq.
>
> Rebecca: *Iqqamukisaaq?*
>
> Lydia: *They are not supposed to be consumed.*
>
> Rebecca: *Yes.*
>
> Lydia: *There are many harmful insects.*
>
> Rebecca: *Do you know insects that cause sickness?*
>
> Lydia: *The* iqqamukisaaq *cause death.*

Traditional Knowledge

Also from Clyde River: Rebecca Hainnu interviewing Aulaqiaq Areak, July 2006

Rebecca: Iqqalukisaaq?

Aulaqiaq: Iqqamukisaaq. *Maybe it is an "M." They walk at the bottom of the water but then they can float, too.*

Rebecca: *Really*, iqqamukisaaq.

Aulaqiaq: Iqqamukisaaq *looks like* minguq. *It looks like they release gas when they come up to the surface of the water.*

Order: Coleoptera

Water Scavenger Beetles
Hydrophilidae

Order: Coleoptera – Beetles; Family: Hydrophilidae – Water Scavenger Beetles

The beetles belonging to this family are commonly known as water scavenger beetles in English.

All water beetles are named *iqqamukisaat* in Igloolik and Clyde River, from the word *iqqaq*, which means "deep in the water." In Kugluktuk, they are simply referred to as *kingok*, or "bugs from the lake." In Baker Lake, elders identified these beetles as *pootoogooqsiut*, while in Iqaluit they can be called *aqammukitaaq*, *ikarmikitaa*, or *imarmiutait*. In all of the communities, elders also simply called them *minguq*, the generic name for any type of beetle.

The beetles in this family are mostly **aquatic scavengers**. About 70% are aquatic, while the other 30% are **terrestrial**.

Description

Adults:
Water scavenger beetles are oval and their backs are **convex** (domed, like an igloo). They can be black, dull green, brown, or yellow, and they range from one to forty millimetres in length. They have short, **clubbed** antennae and their **palps** are usually longer than their antennae. Some species have sharp spines on their undersides. The hind legs are flat and have fringes of hair.

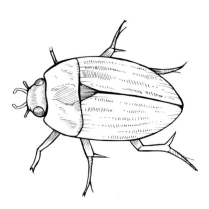

Larvae:
The water scavenger beetle larva is tapered, and has ten **abdominal** segments. It has a hardened head and toothed **mandibles**. The larva has only one claw on its **tarsus** (segment at the end of the leg, sort of like a foot).

Life Cycle

Complete Metamorphosis:
The aquatic hydrophilids construct silky cases in which to deposit their eggs, and then attach these cases to plants in the water. Larvae hatch out of the eggs and go through various **moults** until finally they leave the water and dig holes in the earth, in which they later **pupate**. When the adult emerges from the **pupa**, it heads back to the water, completing the cycle.

Food and Feeding

The adults in this family are mostly scavengers, eating decaying or rotting plant and animal matter.

The larvae, on the other hand, are **predaceous**—they eat all kinds of small aquatic animals.

Habitat

Most species are aquatic and live in quiet streams and ponds. The adults are good at swimming, doing so by kicking their back legs one at a time. They can stay underwater since they contain air in bubbles on the undersides of their bodies. The terrestrial species tend to live near water, in dung, under rocks, or even in moss.

Range

Only a few species of hydrophilids have been found in tundra locations. The species found have all been in the Low Arctic.

Did You Know?

When looking at water beetles in ponds, dytiscids and hydrophilids can be confused one for the other, but there are a couple of quick ways to tell them apart: by how they swim and how they breathe.

The beetles in the family Hydrophilidae swim by kicking their back legs alternately or one at a time, while the dytiscids swim by doing a frog-like kick with both back legs at the same time. As for breathing, beetles in the Dytiscidae family come to the surface hind end first, and attach small air bubbles there. These aquatic beetles were probably the first scuba divers! Hydrophilids, on the other hand, come to the surface head first, use their antennae to break the surface tension, and hold air bubbles on the undersides of their bodies.

The larvae of these two groups are also similar, and require a closer look. Dytiscid larvae have two claws at the end of their tarsi, while hydrophilid larvae only have one. Also, dytiscid larvae have mandibles without teeth, while hydrophilid larvae have mandibles with teeth.

Photo by Tom Murray

Order: *Coleoptera*

Carrion Beetles

Silphidae

Order: Coleoptera – Beetles; Family: Silphidae – Carrion Beetles

The beetles in the family Silphidae are commonly known as carrion beetles in English. No particular Inuktitut name was found for members of this family during the research for this book. However, all beetles can be referred to as *minguq* in Inuktitut.

Carrion beetles are usually quite large and colourful and can be found around the bodies of dead animals. This family is divided into two subfamilies: Silphinae and Nicrophorinae. The species of carrion beetles that likely occur in Nunavut all belong to the Silphinae subfamily and therefore all descriptions that follow relate to that group.

Description

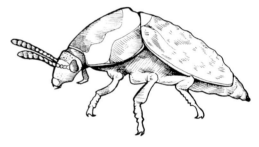

Adults:
Carrion beetles have soft and somewhat flat bodies. They range from three to thirty-five millimetres in length. Their antennae are **clubbed** and have eleven segments. Most species have black **pronota** that are wider than they are long—in fact, they are larger than the beetles' heads. The **elytra** are either black or brown, usually with three ridges, and may be coloured with bright red or orange spots. The elytra almost cover the beetles' **abdomens**.

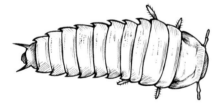

Larvae:
The larvae are soft, mostly white **grubs**. They can have brown, black, or reddish markings on their backs. They have short, three-segmented antennae.

Life Cycle

Complete Metamorphosis:
In the Silphinae subfamily, adults must first locate animal carcasses before they are ready to mate. Once a carcass is found, mating can take place. The females deposit their eggs in the soil around the carcass. In two to seven days, the eggs hatch and the larvae move to the carcass to feed. These beetles have three **instars**, each followed by a **moult**. The first moult takes place within three to seven days. The second and third instars feed for three to ten days each before moulting. Once the final instar is reached, the larvae **pupate** in the soil for fourteen to twenty-one days and then emerge as adults.

Food and Feeding

Both as adults and larvae, silphids feed on **carrion**, the decaying flesh of animals. These beetles have an advanced sense of smell, which helps them to find carcasses. Rarely, they might also feed on dead plants or fungi.

Habitat

Silphids live on the tundra, wherever carrion is found.

Range

There are only two species of carrion beetles found in the Arctic, and they are both found in the Low Arctic.

Did You Know?

Carrion beetles are not the only insects that you will find on animal carcasses. Many fly larvae also feed on carrion. Silphinae carrion beetles avoid competition for food with the fly larvae at animal carcasses by only feeding once the fly larvae have finished. That means the beetle eggs only hatch once the fly larvae are done eating and are ready to pupate.

Photo by Scott Nelson

Photo by Scott Nelson

Photo by Floyd Schrock

Order: Coleoptera

Rove Beetles

Staphylinidae

Order: Coleoptera – Beetles; Family: Staphylinidae – Rove Beetles

The beetles in this family are commonly known as rove beetles in English. An elder from Baker Lake recognized them as *aullak* in Inuktitut.

Description

Adults:
Most adult rove beetles are elongated and slender, with short **elytra**. The elytra generally only cover the first two segments of the abdomen, leaving five or six segments exposed. Rove beetles range from one to twenty-five millimetres in length. They have flexible abdomens, which they sometimes carry like scorpions' stingers. Rove beetles have threadlike, **clubbed** antennae. Their hindwings are membranous and are tucked under the short elytra when the beetles are not in flight. Rove beetles are usually black or brown, and have chewing mouthparts. In fact, their **mandibles** are long and sharp and usually cross at the fronts of their heads.

Larvae:
Rove beetle larvae are slender and have distinct segments. There are ten abdomen segments, the last two being darker and tapering to a rounded end. The head is bigger and darker than the body. These larvae have antennae with three joints.

Life Cycle

Complete Metamorphosis:
The female lays white eggs that hatch into larvae within a few days or weeks. When it is time to **pupate**, the larvae spin silken **cocoons**. All immature stages occur rapidly, allowing the adults to live long lives.

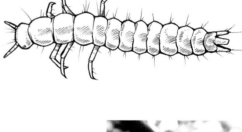

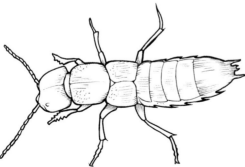

Photo by Johannes F. Skaftason

Food and Feeding

Most of the adults are **predaceous**, feeding on other insects or worms. However, there are some rove beetles that also feed on fungi, dead or decaying plants or animals, and, rarely, live plants.

Some larvae are predaceous, while others are **parasitic**.

Habitat

Rove beetles are strictly **terrestrial**. Because so much of their abdomens are exposed and can dry up rapidly, however, these beetles live in moist environments, such as under leaves. They can easily crawl around in such environments because of their long, flexible bodies. They can also live on plant materials by the sea or near rivers and streams. Some of these beetles also live in animal faeces, under rocks, on dead animals, and even in the nests of some animals.

Range

As with most insects in the Arctic, many of the rove beetle species have not yet been identified. So far, about fifteen species of rove beetles have been found in the Arctic, but it is assumed that at least that many more are yet to be discovered. Only two species have been reported in the High Arctic.

Photo by Johannes F. Skaftason

Did You Know?

Some beetles in the Staphylinidae family have glands that produce defensive substances—smells that can ward off possible enemies. One group of these beetles also produce a substance that enables them to walk over the surface of freshwater (if they happen to fall in) and make their way back to dry land.

Another group of rove beetles, not found in Nunavut, produces one of the most powerful toxins of any animal. If human skin is exposed to the toxin, a rash ensues, and this same toxin can do severe damage to our eyes. This toxin has been used to heal chronic lesions, and to help cure some cancerous growths.

Order: *Coleoptera*

Click Beetles

Elateridae

Order: Coleoptera – Beetles

Family: Elateridae – Click Beetles

The beetles in this family are commonly known as click beetles in English. In the Iqaluit area, as with some of the other beetles, click beetles are simply called *minguq* in Inuktitut. An elder from Taloyoak identified this beetle as *qauguklik*. In Kugluktuk, elders recognized these beetles as "the ones that make the clicking noise," or *bigligiak*.

Description

Adults:

Click beetles have long, flat bodies that are rectangular but rounded at both ends. They have short legs. They can be between 0.9 and 102 millimetres in length, but most range from twelve to thirty millimetres. They are usually black or brown and have chewing mouthparts with **mandibles**. The **labrum** (upper lip) is exposed. These beetles have serrated (like the blade of some knives) antennae. Their **pronota** are pointed at the back corners.

The click beetle is rather unique due to certain features. Its **thorax** is made up of three segments, which is always the case in beetles, but in the click beetle the first two segments are only loosely joined. It also has a long spine on the underside of its body that fits into a groove. When the spine is inserted into the groove, the beetle flips through the air, always landing on its feet. When this happens, a clicking sound is made.

Larvae:

Click beetle larvae are called wireworms. They are elongated, and can be round to flat. They have hardened **exoskeletons**, strong mandibles, and thin legs near their heads. They are shiny and can be yellow to brownish-yellow, dark brown, or even black.

Life Cycle

Complete Metamorphosis:
Click beetles live longer than most other insects. Adults lay tiny white eggs in the soil early in the spring. They die soon after laying eggs. In a few days or weeks, the eggs hatch and the wireworms start to feed. They are mature after two to six years. In the middle of summer, once mature, they **pupate** in underground cells. After the pupal stage, the adults stay underground until the following spring.

Photo by Johannes F. Skaftason

Food and Feeding

Adult click beetles eat plants. Most larvae feed on roots or seeds, although a few species are **predaceous**. However, it is thought that in the Arctic species found so far, both the adults and the larvae eat plants.

Habitat

Adults live on plants, while larvae live in the soil or under rotting wood.

Range

Very few species of click beetles have been found in the Low Arctic. None has been found in the High Arctic.

Did You Know?

Bioluminescence is a living organism's ability to produce light. Many creatures have been described as being bioluminescent. Insects with organs that produce light occur in Collembola (primitive bugs), Hemiptera (true bugs), Coleoptera (beetles), and Diptera (flies). Insect bioluminescence has developed so that insects of the same species can signal to their mates at night. One of the largest populations of such beetles is the click beetle family in Jamaica. In Nunavut, this type of adaptation would not be advantageous, as there is almost twenty-four hour sunshine throughout the territory during the mating season.

Order: *Coleoptera*

Ladybugs

Coccinelidae

Order: Coleoptera – Beetles

Family: Coccinelidae – Ladybird Beetles, Ladybugs

Insects of the Coccinelidae family are commonly known as ladybugs or ladybird beetles in English. No particular Inuktitut name was found for this family during the research for this book, but all beetles can be referred to as *minguq* in Inuktitut. Insects of the Coccinelidae family are some of the most easily recognized because of their bright colours and spots.

Description

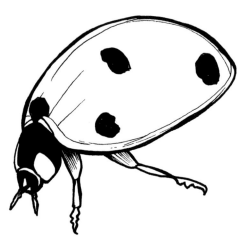

Adults:
Ladybugs are round or oval-shaped. They range from 0.3 to ten millimetres in length. Their **elytra** cover the abdomen and meet all the way down the middle. The hindwings are membranous. The head is partly hidden by the **pronotum** and can retract under it. As well, the ladybug's short legs can also retract under its body. Ladybugs often do this to simulate being dead, as a defence mechanism.

The ladybug's segmented, **clubbed** antennae are shorter than its body. The upper side of its body is **convex** and the underside is flat. These beetles are brightly coloured and often have spots.

Larvae:
Ladybug larvae are soft-bodied, long, and flat. They have six legs. They are covered with spines, and can be spotted or banded.

Life Cycle

Adult ladybugs **overwinter** in dry, sheltered areas. In the spring, once the weather is warm enough, breeding begins. Females can lay as many as one thousand eggs at once. The eggs take approximately four days to hatch. The larva **moults** three times, and after about three months it forms a **pupa**. After about one week as a pupa, the skin along its back splits open so that the adult can emerge. The adults feed for about one week before finding places to overwinter. Then the cycle begins again.

Food and Feeding

Both the adults and the larvae feed on aphids and other small insects.

Habitat

Ladybugs are found most often near their favourite food—aphids. Aphids feed on tundra plants, so you might observe ladybugs on willows, mountain avens, or chickweed.

Range

Species from the Coccinelidae family have been reported in the Low Arctic and also in Kugluktuk.

Did You Know?

One of the common names for this beetle, "ladybird beetle," has an interesting origin. In Europe, the "lady" in ladybird refers to the Virgin Mary. The red colour of the hard wings represents her cloak, while the black spots represent her joys and sorrows.

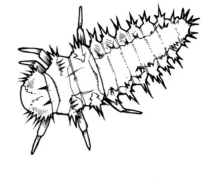

Photo by S.A. Marshall

Traditional Knowledge

Allen Niptanatiak, an elder in Kugluktuk, recalls that they used to see more ladybugs than they do now. When he saw one last summer, he was happy to note that ladybugs might be coming back.

Weevils

Curculionidae

Order: Coleoptera – Beetles; Family: Curculionidae – Weevils

The beetles in this family are commonly known as weevils in English. No particular Inuktitut name was found for this family during the research for this book. However, all beetles can be referred to as *minguq* in Inuktitut. Weevils are also referred to as snout beetles because they have long snouts, similar to those of elephants.

Description

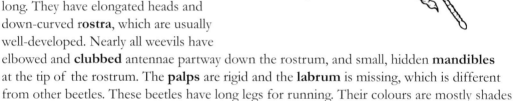

Adults:
Weevil adults have hard bodies and they can measure from one to forty millimetres long. They have elongated heads and down-curved **rostra**, which are usually well-developed. Nearly all weevils have elbowed and **clubbed** antennae partway down the rostrum, and small, hidden **mandibles** at the tip of the rostrum. The **palps** are rigid and the **labrum** is missing, which is different from other beetles. These beetles have long legs for running. Their colours are mostly shades of brown that blend in with bark, rocks, or dirt, so they can stay disguised when they are moving.

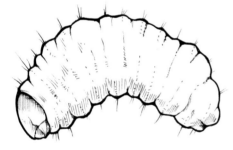

Larvae:
The larvae are short, thick-bodied, fleshy **grubs** and are pale, white, or creamy in colour. They are usually C-shaped, and they are **apodous**, which means that they do not have feet or foot-like members.

Life Cycle

Weevil eggs are elongated, and are white or cream in colour. Some weevil females use their long rostra as drills to make holes in fruit, seeds, or stems, into which they then lay their eggs. Other females simply lay their eggs on plants. Either way, when the larvae hatch, there is plenty of food for them since they are **herbivores**.

Some weevil larvae continue to eat until it is time to **overwinter**. Then they **pupate** in the spring and become adults. Other weevil larvae pupate during the same summer in which they hatch, leaving the adults to overwinter. Whether they overwinter as larvae or adults, all weevils spend the winter under leaves or perhaps under rocks, where they are protected from the cold and wind.

Food and Feeding

Weevils are herbivores. Both the adults and the larvae feed on all parts of plants. In the South, they can be pests, consuming agricultural crops and forests.

Habitat

Most weevils feed specifically on one type of plant, and therefore live on and around that plant. In Nunavut, not enough research has been done to identify which plants weevils feed on. A small sample studied personally by the author chose to eat willow leaves over all other vegetation.

Photo by Carolyn Mallory

Habitat

Over a dozen species of weevils have been found in the Arctic, although quite a number of them are found in forested areas and therefore not in Nunavut. A few species have been found in the Low Arctic regions of Nunavut, while only one species has been found in the High Arctic.

Did You Know?

If discovered, weevils usually tuck their legs up under their bodies and drop, effectively playing dead. In Iqaluit at the end of May, you can find weevils by overturning rocks and looking carefully at their undersides. The weevils are coloured like the rocks and will stay very still.

Photo by Kathy Thornhill

Flies

Order: Diptera – Flies

The generic Inuktitut name for flies in the Igloolik, Iqaluit, and Clyde River areas is *ananngiq*. In Baker Lake, flies are more often referred to as *niviuvak*. The Inuktitut word for the **maggot** (fly larva) is *qitirulliq*. Specific types of flies often have other names as well.

Flies are the most numerous insects in the Arctic. They are divided into two suborders: Nematocera (long-horned flies) and Brachycera (short-horned flies). Here is a table that compares their features.

Nematocera	Brachycera
Small, slender, long-legged	Robust, stocky
Long, slender antennae	Short antennae (less than eight segments)
Look like mosquitoes or midges	Look like houseflies
Larvae have well-developed heads (except for Cecidomyiidae)	Larvae have no distinct heads
Larvae's mouthparts move horizontally	Larvae's mouthparts move vertically
Most are **aquatic**	Most are **terrestrial**
Larvae go through four **instars** (stages of growth)	Larvae go through three instars
Adults feed on nectar and may need blood meals to produce young	Adults are **predators** or **scavengers**

Description

Adults:
Flies can be anywhere from 0.5 to five millimetres long and are soft-bodied. Each has an enlarged **thorax** to accommodate its relatively large wing muscles, while the other two body segments—the head and the abdomen—are reduced in size. Adult flies have only one pair of membranous wings. The hindwings are reduced to knobby structures known as **halteres**, which are used for balance when flying. The antennae are variable, and depend on the suborder. Flies have large **compound eyes** and well-developed **palps**. They also have piercing, lapping, or sucking mouthparts.

Larvae:
Fly larvae, known as maggots, are legless and wormlike. They are pale in colour. Some have distinct heads, while others do not.

Life Cycle

Complete Metamorphosis:
Female flies lay their eggs in moist, protected areas, such as in soil, within plants, on other animals, or in aquatic habitats. The maggots hatch and then feed until they are large enough to **pupate**. From the pupae, the adults emerge and begin the task of finding mates so that the cycle can begin again.

Food and Feeding

Photo by Anna Ziegler

There is great variety in the feeding habits of adults. Some are **predaceous**, while others feed on plant nectar. Still others are external **parasites** and feed on the blood of **host** organisms. There are also flies that feed on decomposing organic matter. The larvae consume a great variety of different foods as well. Like the adults, some larvae feed on other insects, vertebrates, and mollusks, some feed on faeces, some feed on plants or dead organic matter, and some live as parasites.

Habitat

Flies and their larvae are found in almost every habitat imaginable. Larvae can live in water, inside plants, in dung, or in **carrion**—any place where there is enough moisture to ensure that they do not dry out. Adults are terrestrial but may live near water.

Range

In Arctic North America, thirty-nine families of flies have been identified. A few of these families must live where there are trees and will therefore not be found in Nunavut. Twenty-one families have made it to the High Arctic, mostly from the suborder Nematocera (groups such as Chironimidae in particular).

Did You Know?

With only two wings, flies are some of the best pilots in the world. They can fly backwards, hover, turn in one spot, and even fly upside down!

Traditional Knowledge

From Taloyoak: Ellen Ittunga interviewing Mary Ittunga, July 2006
Mary explained that the *niviuvaks* lay their larvae on meat and fish that are being dried. They lay them when the flesh is still moist and they tend to lay them in between open slits. She said that this is why you have to open up the slits when you're eating dried meat or fish, because the *niviuvak*, like the rat, is considered to be an animal that carries diseases.

Order: *Diptera*

Winter Crane Flies

Trichoceridae

Order: Dipthera – Flies, suborder Nematocera; Family: Trichoceridae – Winter Crane Flies

Members of the Trichoceridae family are commonly known as winter crane flies in English. No specific Inuktitut names were found for this family during the research for this book. Please see the Diptera account for general names for flies. Also, if you know of names for any species in this family, please contact the Nunavut Teaching and Learning Centre.

Winter crane flies are active in the late fall and early spring, and even sometimes in the winter, thus explaining their common English name.

Description

Adults:
Winter crane flies have two to three **ocelli**, which distinguish them from the Tipulidae family (other crane flies). They are slender, long-legged insects, and have long antennae with eight to sixteen segments. Their lower **calypters** (lobes at the bases of their wings) are much reduced or absent.

Larvae:
The winter crane fly larvae are soft-bodied and cylindrical. They possess both lungs and **gills**. They have eleven body segments with exposed head capsules. On each insect, the last segment ends in four lobes. They have short antennae and small eyespots above the base of the jaw.

Life Cycle

Complete Metamorphosis:
Females choose suitable spots to lay their eggs, preferring decaying leaves, manure, or fungus. After a few days, the larvae emerge. They go through four **instars**, before finally becoming **pupae**. From the **pupal stage**, the adult flies emerge and the cycle begins again.

Food and Feeding

Little is known about the feeding habits of adult winter crane flies. The larvae eat decaying vegetable matter or animal droppings.

Habitat

The adults can be found resting inside caves, hollow logs, or similar dark places when they are not flying. Larvae are found in decaying leaves and other moist habitats. In warmer, more southern climates, these flies can be seen on sunny winter days, swarming over bushes and hills.

Range

About three winter crane fly species have been found in the Low Arctic, while only two species have been found in the High Arctic.

Did You Know?

Sunny and warm winter days increase winter crane fly activity, while cold weather lowers their numbers.

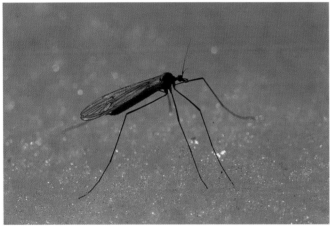

Photo by James Lindsey

Order: Diptera

Crane Flies

Tipulidae

Order: Diptera – Flies, suborder Nematocera; Family: Tipulidae – Crane Flies

Members of the Tipulidae family are commonly known as crane flies in English. In Inuktitut, crane flies are called *tuktuujaq* in Clyde River, Igloolik, and Iqaluit. In Clyde River, they are also called *suluqtuq*. The Inuktitut names for crane flies in other areas are similar. For example, in Kugluktuk they are called *tuktuuyak*. In Taloyoak, the word is still close: *tukturruk*. In Baker Lake, the Inuktitut term for crane flies is *tukturjuk* or *uliniu*.

There are about fourteen thousand species of crane flies worldwide, making it the largest family in the Diptera order.

Description

Photo by Kathy Thornhill

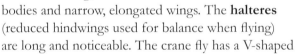

Adult:
Crane flies are either grey or brown, and look like giant mosquitoes. They have very long legs that break off easily when you capture them. They are between eight and sixty millimetres long, with slender bodies and narrow, elongated wings. The **halteres** (reduced hindwings used for balance when flying) are long and noticeable. The crane fly has a V-shaped groove on the top of its **thorax**, large **compound eyes**, and short antennae.

Larvae:
The crane fly larva is also called a "leather jacket" because its skin is tough. It is legless and has a reduced head, which is partly retracted into its body. The head capsule is well-hardened. The larva's body is elongated and its abdomen may be smooth or have fine hairs. It has one pair of breathing pores, surrounded by fleshy lobes.

Photo by Johannes F. Skaftason

Life Cycle

Complete Metamorphosis:
Once the crane fly eggs are laid—in water or wet soil, depending on the genus—they hatch in less than two weeks. Crane flies go through four **moults** before they **pupate**. The **pupal stage**, during which the insect rests and transforms into the adult, is between five and twelve days long. Crane flies can **overwinter** as pupae. Females emerge already containing eggs, ready for fertilization. The adults are short-lived.

Nunavut is at the northern limit of where crane flies can live, and their life cycles may take as long as five years there. If conditions are not good one summer and the crane fly is unable to moult as many times as it needs to, it will overwinter as a larva and continue its growth the next season.

Food and Feeding

Not a lot is known about the adult feeding habits of crane flies. It is thought that some species might feed on the nectar of plants, while others likely do not eat at all. The larvae feed mostly on decaying plant material, though some may feed on living plant tissues. A few species are **predaceous**.

Habitat

Adults can be found in low vegetation, often near water. Larvae are found in freshwater or on moist soil.

Range

Four species of crane flies are known to occur in the High Arctic, and another ten species have been found in the Low Arctic.

Did You Know?

You can tell what sex a crane fly is by the way it flies. Females have a straight and steady way of flying, while males tend to rotate and fly up and down in a wave pattern.

Traditional Knowledge

Norman Attungala, interviewed in Baker Lake, said that he was told that crane flies keep the water surface clean. He also stated that they eat other bugs.

Peter Kunilusie, from Clyde River, described these flies as "the ones that look like they are being juggled." He also said that they are known as "eaters of mosquito brains," and that he had seen crane flies kill mosquitoes on numerous occasions.

The word *tuktunjaq* means "resembles a caribou." This is thought to be the case because both caribou (*tuktu* in Inuktitut) and crane flies have long legs.

Mosquitoes

Culicidae

Order: Diptera, suborder Nematocera; Family: Culicidae – Mosquitoes

Members of the Culicidae family are commonly known as mosquitoes in English. In Inuktitut, they are generally called *qikturiaq* (there are many variations of the spelling). In Kugluktuk, mosquitoes are called *kiktogiak* or *kaligoalik*.

Description

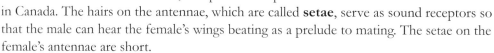

Adults:
Mosquitoes range from three to fifteen millimetres. Adults have small, roundish heads and **compound eyes**. They have long, slender **proboscises** (extended, beak-like mouthparts). The females' mouthparts are modified to allow them to suck blood. Mosquitoes also have long, slender antennae with fifteen segments. Males have feather-like antennae, except for one species in Canada. The hairs on the antennae, which are called **setae**, serve as sound receptors so that the male can hear the female's wings beating as a prelude to mating. The setae on the female's antennae are short.

The mosquito's **thorax** is egg-shaped and, in Arctic species, there is additional hair on this body segment. The **abdomen** has ten segments, also covered with hair or **scales**. The abdomen is taller than it is wide, and is somewhat flat. Adult mosquitoes have two elongated, narrow wings that lie flat on the abdomen when the mosquito is at rest. The wing veins are densely scaled. The legs are also covered in scales—either fully concealed with dark ones or banded with light ones so as to seem striped.

Larvae:
Mosquito larvae are called wrigglers and are **aquatic**. They are slender-bodied and legless, and they swim rapidly by lashing their bodies from side to side. To breathe, they have respiratory **syphons** that they can send up to the surface of the water for air. They have brushes on the upper lips of their mouths that help them gather food.

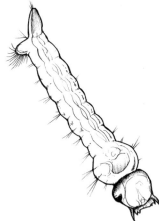

Pupae:
The **pupae** are aquatic and look like tiny tadpoles (baby frogs). They swim freely and rapidly, which is something the pupae of

most other insects do not do. When they are disturbed, they swim to the bottom of the puddle or pond and hide.

Life Cycle

Complete Metamorphosis:
Mosquito eggs are laid in water or moist soil where temporary pools of water are likely to collect. The eggs need standing water (water that is not running) to hatch. The mosquitoes go through four **instars** after which they **pupate**.

Photo by Carolyn Mallory

When the adult is ready to emerge, the pupa makes its way to the surface of the water. The thorax ruptures along the mid-back, allowing the adult to free itself of the pupa. The males usually emerge a few days before the females. They swarm together soon after emergence and, when a female enters the swarm, mating occurs. Females mate once, while the males continue to swarm. The females then seek blood meals to initiate and develop new batches of eggs. Once the blood meals are ingested, the females must deposit their eggs. The females are inactive as the eggs develop.

The *Aedes* genus of mosquitoes, the most common in the Arctic, **overwinter** as eggs. Two *Culiseta* genera overwinter as inseminated females who must seek blood meals first thing in the spring. Mosquitoes live for less than one month as adults.

Food and Feeding

Adult mosquitoes feed on nectar and plant juices. The blood meals are only necessary for females, in order to help with egg development. The larvae eat fine organic materials in the water, as well as algae and protozoans. They are filter feeders. They use the brushes around their mouths to collect food and direct it into their mouths.

Habitat

The larvae and pupae are aquatic. Their habitats can be anything from temporary snow-melt pools to marshes and even water in discarded containers—they simply need standing water. The adults are found everywhere, in all kinds of environments.

Range

About twenty-five species of mosquitoes are found in the Arctic. They can live everywhere in Nunavut, except on a few of the small Arctic Islands. That being said, only two species, *Aedes impiger* and *Aedes nigripes*, breed in the High Arctic.

Did You Know?

When female mosquitoes are looking for blood meals, they can detect carbon dioxide (the gas that is expelled when animals breathe) in the air, and follow the scent to reach animals on which they can feed.

Traditional Knowledge

From Iqaluit: Sylvia Inuaraq interviewing Celestine Erkidjuk, July 2006

When I lived in Kivalliq, in order to be outside and be able to breathe some air, we had to burn what we call urju. *This prevented us [from] breathing in mosquitoes.*

From Kugluktuk: Annie Kellogok interviewing Joseph Niptanatiak, July 2006

In the olden days, there was a lot of kiktogiaks. *It was good for the hunters because they only had bows and arrows. When the bugs were swarming around the caribou, the caribou would run and get close to the hunters—close enough for bows and arrows.*

From Taloyoak: Ellen Ittunga interviewing Bernadette Uttaq, July 2006

There's a story and pihiq *[song] about the* niviuvak *and* kitturiaq. *These two insects were very competitive towards each other. The* kitturiaq *challenged the* niviuvak *by taunting him, saying, "You don't stand a chance with me because you have no poker."*
The niviuvak *accepted the* kitturiaq's *challenge by replying, as calmly as possible, "I can probably win even though I don't have a poker."*
The insects went into battle and the kitturiaq *tried his best to poke the* niviuvak, *but he couldn't poke the insect. He kept missing. The* niviuvak *was so swift that he got hold of the* kitturiaq *and threw him across [the room]. The* kitturiaq *started to cry.*

The pihiq goes like this:

ᑐᖃᓗᐅᐳᓂ ᐊᑭᓚᖅᔭᖕᓗ ᐊᖃᐅᐲᕐᓯᖕᒥᕐᐳ
ᖅᓱᓕᕐᓯᓚᓗᖕᔭᑦ ᑐᓄᖕᖠᖅᓯᑦᓴ ᖃᐃᓕᓴ ᑕᒪ
ᓯᖁᐃ ᓯᖁ ᑳᖁ ᖕᒪ ᐊᖁ
ᑐᖃᖕᕋᓄᐊᓗ ᖃ ᐊᑭᓕᖕᒥᑎᖕ ᐊᓚᖃᑎᕋᓗᖃᐃᐳᓗ
ᖅᓱᐊᐳᒥᕐᓗᖕᒪ ᑐᓄᖕᖠᖅᓯᑦᑐ ᖃᐃᓕᓴ ᑕᒪ
ᓯᖁᐃ ᓯᖁ ᓯᖁ ᐊᓯᖁ ᐊᖁ

Traditional Knowledge

From Clyde River: Rebecca Hainu interviewing Kalluk Palituq, July 2006

Mosquitoes draw so much blood from caribou that they can kill them, and also dogs. That is how it used to be before, but it's been a long time since that happened.

From Baker Lake: Brenda Qiyuk inteviewing Silas Aittauq, July 2006

We knew caribou were close by when there were many mosquitoes around.

Knud Rasmussen collected stories and myths that were published in 1929 in *Intellectual Culture of the Iglulik Eskimos*. This story about how the mosquito first came to be was published in that volume.

How the Mosquitoes First Came:

There was once a village where people were dying of starvation. At last there were only two women left alive, and they managed to exist by eating each other's lice. When all the rest were dead, they left their village and tried to save their lives. They reached the dwellings of men, and told them how they had kept themselves alive simply by eating lice. But no one in that village would believe what they said, thinking rather that they must have lived on the dead bodies of their neighbours. And, thinking this to be the case, they killed the two women. They killed them and cut them open to see what was inside them; and lo, not a single scrap of human flesh was there in the stomachs— they were full of lice. But then all the lice suddenly came to life, and this time they had wings, and flew out of the bellies of the dead women and darkened the sky.

Thus mosquitoes first came.

Told by Inugpasugjuk

Order: *Diptera*

65

Black Flies

Simuliidae

Order: Diptera – Flies, suborder Nematocera; Family: Simuliidae – Black Flies

Members of the Simuliidae family are commonly known as black flies in English. In Kugluktuk, black flies are called by the generic Inuktitut name for any fly, *niviovak*. In Baker Lake, it is much the same, as black flies are referred to as *niviuvak*. However, they can also be called *anangiq* (another typical fly name) or *milugiaq*. In Clyde River, elders refer to the black fly as *qaumajaq* and *anangiq*. In Taloyoak, they are again called *niviuvak*. In Iqaluit, black flies are called *anangiq* or *kumaujaq*.

Description

Adults:
Black flies are small, squat flies with curved **thoraces** that make them look humpbacked. These grey or black flies are two to five millimetres long, and have large, rounded wings. The veins at the front of the wings are very pronounced. Adults have short, beaded antennae. The large compound eyes on the male are almost connected, while the eyes on the female are completely separate. The **proboscis** is short and thick; the female's is adapted to suck blood. These small flies have short legs and elongated **abdomens** with eleven segments.

Larvae:
Black fly larvae are elongated and swollen at their hind ends. They range from five to fifteen millimetres in length and can vary in colour from pale to dark brown. These larvae have well-developed, heavily **sclerotized** (hardened) heads. They have short antennae with only four segments. On their heads, they also have fans or brushes near their mouths to help them feed. Their abdomens have eight segments and one **proleg**. They each have a series of hooks at the ends of their abdomens.

Pupae:
The **pupae** of black flies are said to be **obtect**, which means that the legs, wings, and antennae are all glued to the body inside the pupa. On top of the pupae are strangely shaped gills, used for breathing underwater. The **cocoons** of black fly pupae vary greatly. Some have hardly any cocoon, while others have sleeves that cover part or all of the pupae.

Life Cycle

Black flies need flowing water to live. Most females drop their eggs into water as they are flying low. Some, however, will also lay their eggs on vegetation in the water. Larvae hatch anywhere from a few days to many months after the eggs are laid, depending on whether they overwinter as eggs or not. Females typically lay between 150 and 450 eggs. The first

instar has a spine on its head that allows it to escape from the egg. Black fly larvae go through six to nine instars. The last instar spins a cocoon that will protect and anchor the pupa. The pupal stage typically lasts from four to seven days. The adult emerges from the pupa through a slit in the top. As the fly emerges, it is surrounded by an air bubble, in which it floats to the surface of the water. The bubble pops and the adult flies away. It immediately goes to rest so that its body can harden. After her body has hardened, the new female is ready to mate. In most Arctic black flies, the female produces eggs that will overwinter. In some Arctic black flies, a male is not needed for reproduction. The female can reproduce all on her own.

Food and Feeding

The adults feed mostly on plant nectar. The female's proboscis is also made to cut through skin to have a blood meal, though, because blood meals are necessary for egg development in most species. However, in the Arctic, many species of black flies have adapted so that they do not need blood meals to successfully produce eggs. This is likely because animals from which the flies could extract blood are not always easy to find. The larvae attach themselves to rocks or plants under the water with silk string, using tiny hooks at the ends of their abdomens. Once they are attached, they use fans near their mouths to catch small organic particles, algae, and bacteria that float by.

Habitat

The eggs are laid in moving water, and the larvae grow there. As for the adults, they can live almost anywhere, as long as they return to running water to lay their eggs.

Photo by James Lindsey

Range

Just over forty species of black flies have been reported in Nunavut. In each of Repulse Bay, Resolute Bay, Igloolik, and Whale Cove, one species has been reported. Between two to three species have been reported around each of the following communities: Iqaluit, Pangnirtung, Clyde River, and Kimmirut. In each of Kugluktuk, Chesterfield Inlet, Taloyoak, Cambridge Bay, and Coral Harbour, four or five different species have been found. About seven species of black flies have been reported from the Rankin Inlet area. Arviat and Baker Lake have the most species of black flies in the territory, with Arviat having about sixteen species and Baker Lake having about twenty-seven species. That means that almost all of the black fly species that live in Nunavut can be found in the Baker Lake area.

Did You Know?

Black flies are very influenced by colour. They like dark colours better than pale ones and they prefer blue, purple, brown, and black to white and yellow. If you are going to be outside where there are many black flies, you should dress in colours that they like less, hopefully making you less attractive to them.

No-see-ums

Ceratopogonidae

Order: Diptera – Flies, suborder Nematocera

Family: Ceratopogonidae – No-see-ums, Sand Flies

Members of the Ceratopogonidae family are commonly known as no-see-ums or sand flies in English. No specific Inuktitut names were found for this family during the research for this book. Please see the Diptera account for general names for flies. Also, if you know of any names for species in this family, please contact the Nunavut Teaching and Learning Centre.

No-see-ums are very tiny—they are the smallest of the biting flies.

Description

Adults:
No-see-ums range from one to six millimetres long. They are thin, with small, rounded heads. In males, the antennae are hairy and feather-like. Their **proboscises** are about the same length as their heads. The females' mouthparts are adapted for piercing. Their **thoraces** are oval-shaped. These insects fold their wings one on top of the other when they are at rest, and the male's wings are narrower than the female's. The **abdomen** has ten segments.

Larvae:
The no-see-um larvae are tiny, long, and thin. They have strong jaws and their heads are usually **sclerotized**.

Life Cycle

Complete Metamorphosis:
The female lays eggs in **gelatinous** (jelly-like) masses in ponds or streams. Once the eggs hatch, the larvae live in damp places and go through three to four **instars**. They **pupate** either in damp locations or in water, where they continue their development. The following spring, the adults emerge, mate, and begin the cycle again.

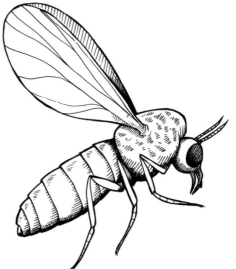

Food and Feeding

Adults feed on nectar. Females need blood meals for egg development; however, it is thought that most northern species are **autogenous**, which means that they can make their eggs without the protein from a blood meal.

The feeding habits of no-see-um larvae depend on whether they are **herbivores** or **carnivores**. The herbivores live mostly on the ground, where it is moist. They eat algae, fungi, or spores. The carnivorous larvae may live in the soil or in the water, where they feed on larger insect larvae.

Habitat

No-see-um adults do not generally move very far from where they develop as larvae. Adults and larvae are found in most moist areas, typically near streams and rivers, or on the shores of lakes or other marine areas.

Range

About six species have been reported in the Arctic. These occur both on the Arctic islands and on the mainland.

Did You Know?

Some species of no-see-ums are important **pollinators** of tropical plants. For example, no-see-ums pollinate cacao flowers. So next time you have a bite of chocolate, take a minute to appreciate the tiny insects that helped in its production.

Order: *Diptera*

Non-biting Midges

Chironomidae

Order: Diptera – Flies, suborder Nematocera; Family: Chironomidae – Non-biting Midges

Chironomidae are commonly known as non-biting midges in English. In Iqaluit, non-biting midges are called by many Inuktitut names: *niviarjuit*, *qitturiakallai*, and *niviuvait*. In Kugluktuk, they are called *nivik*. In Taloyoak, non-biting midges are called *kittusalik*. In Baker Lake, these midges are called *niviaqyuk*, *umiaqjuk*, and *milugiaq*.

Non-biting midges are the dominant group of insects in the Arctic.

Description

Adults:
Non-biting midges look like mosquitoes, but they are more delicate and do not have the long **proboscises** of mosquitoes. They range from one to ten millimetres long and are soft-bodied. Non-biting midges are generally black or brown, with long, narrow wings that are held out to the sides when the insect is at rest. They have reduced mouthparts. Males have feather-like antennae, while females have much less hairy antennae. Chironomids have **compound eyes**.

Larvae:
The larvae are slender, long, and curved. They have hard head capsules and two sets of **prolegs**—one pair behind the head and the other at the hind end. Some larvae have red **hemoglobin** in their bodies and therefore are red in colour. These larvae are called bloodworms.

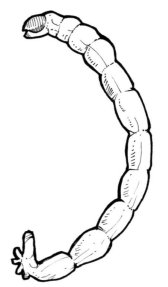

Life Cycle

Complete Metamorphosis:
The female midge lays eggs in water or on vegetation. The eggs can be single or in a jelly-like mass. Once the larvae hatch, they go through four **moults**. This can last from two weeks to four years. Circumpolar species may take even longer to reach adulthood. In freezing conditions, the larvae curl up in **cocoons** to **overwinter**. After four **instars**, the larvae **pupate**. This resting stage only lasts a few days. The adults often emerge in huge numbers. They then form mating clouds, and the cycle begins again.

Food and Feeding

Adult non-biting midges drink water and maybe nectar, but it is generally thought that they do not feed. As for the larvae, some feed on **micro-organisms** (tiny critters that we cannot see with the naked eye) or dead leaves, while others may be **predators** or even **parasites**, living on other organisms and contributing nothing in return.

Habitat

Adult midges are found near rivers, streams, ponds, and lakes. Some varieties are more **terrestrial**, and are thus not found near water. Some midges are even **marine**, meaning that they live by the sea. The larvae can be **aquatic** and may live in all kinds of bodies of water. They can also be found in fine sand or gravel, or under dung.

Photo by Johannes F. Skaftason

Range

Over 140 species of midges have been found in the North American Arctic. They are the predominant group of insects in the High Arctic, with at least sixty-six species having been identified.

Did You Know?

Chironomids occur in very large groups—one hundred thousand per square metre is not unheard of. Their huge numbers make them a very important part of the food chain. They are eaten by birds, fish, and other insects.

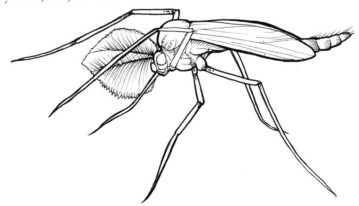

Order: *Diptera*

Dark-winged Fungus Gnats

Sciaridae

Order: Diptera – Flies, suborder Nematocera

Family: Sciaridae – Dark-winged Fungus Gnats

Members of the Sciaridae family are commonly known as dark-winged fungus gnats in English. No specific Inuktitut names were found for this family during the research for this book. Please see the Diptera account for general names for flies. Also, if you know of names for any species in this family, please contact the Nunavut Teaching and Learning Centre.

The dark-winged fungus gnat family is one of the least studied groups in the Diptera order.

Description

Adults:
Dark-winged fungus gnats are quite small, measuring between one and eleven millimetres long. They can be black, yellow, or brown and, like all flies, they have **compound eyes**. Their large eyes meet above the base of the antennae. They also have three **ocelli**. Their heads are small and rounded, with antennae that have between eight and sixteen segments. Fungus gnats have dark, slender bodies with dusky, dark wings and long legs. Females of some species are wingless.

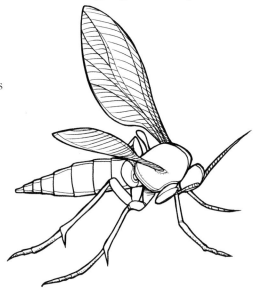

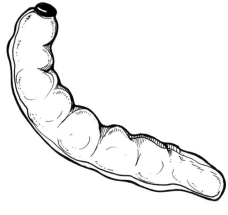

Larvae:
The dark-winged fungus gnat's larvae are legless, white, wormlike creatures with shiny black heads. They have smooth, semi-transparent skin and are about six millimetres long. Their heads are **sclerotized**, with small antennae on top and comb-like brushes underneath. Their bodies have twelve segments.

Life Cycle

Complete Metamorphosis:
There is not much known about the life cycles of these flies. The larvae go through four **instars**. The females of some species can predetermine the sex of their offspring. The adults only live for about two weeks.

Food and Feeding

Adult fungus gnats feed on fungi and decaying organic matter. Some also feed on healthy plants. Larvae feed on decaying plant material as well as animal faeces and fungi.

Habitat

These flies like humid environments. Although the larvae are terrestrial, they prefer the moist environments found under rotting vegetable matter.

Range

Fourteen species of fungus gnats have been reported on the tundra. These flies are not well-known, and likely there are just as many species that have not been described yet. Most of the species found thus far have been in the High Arctic.

Photo by Peter Bryant

Did You Know?

Some plants use scents or even their flower structures to trick male fungus gnats into thinking that they are female gnats. Thus, the male copulates with the flower, pollinating the plant.

Gall Midges

Cecidomyiidae

Suborder: Nematocera   Family: Cecidomyiidae – Gall Midges

Members of the Cecidomyiidae family are commonly known as gall midges in English. No specific Inuktitut names were found for this family during the research for this book. Please see the Diptera account for general names for flies. Also, if you know of names for species in this family, please contact the Nunavut Teaching and Learning Centre.

Description

Adults:
Gall midges are tiny flies—less than three millimetres long. They are slender, with elongated legs and antennae. The **compound eyes** on gall midges meet above the antennae. Gall midges have reduced mouthparts and hairy, sparsely veined wings.

Larvae:
Gall midge larvae are small **maggots** without legs. They have hardened, dark, cone-shaped head capsules. Their mouthparts are reduced and modified for liquid diets. Their antennae have two segments. Many of the maggots are brightly coloured (red, orange, pink, or yellow).

Life Cycle

Complete Metamorphosis:
The females lay about one hundred eggs over their short lives, which are only a few days long. The eggs hatch quickly and the larvae begin to feed. In a great many species, the larvae form **galls** in which they eat and live. A gall occurs when plant cells are altered by a chemical injected by the insect. This allows the larva (one or several) to live inside this plant construction. Once the larva matures, it **pupates**. Oftentimes, the pupae **overwinter** and the adults emerge in the spring to begin the cycle again.

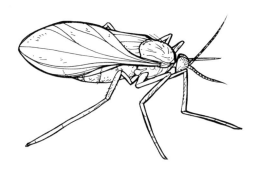

Food and Feeding

About two-thirds of gall midge larvae cause galls to form on plants, and then feed inside these galls. Other larvae may feed on plants without forming galls, eating decaying organic matter or fungi. There are even a few larvae that are **predaceous**, feeding on small insects. Adult gall midges do not feed.

Habitat

Gall midges live in a wide variety of habitats. When inhabiting a gall, they live in plants. One or several larvae can inhabit the same gall, depending on the species. Some midges live in decaying organic matter, in fungi, or under bark. The adults live around the same habitats, as this is where they will eventually lay their eggs. One of the High Arctic species creates galls on willow shrubs.

Photo by Johannes F. Skaftason

Range

A few species of Cecidomyiidae have been identified in the Arctic so far. Only two species reach the High Arctic.

Did You Know?

There are some species of gall midges in which daughter larvae are sometimes produced inside mother larvae. They eventually eat their way out of these mothers, killing them. This type of reproduction can occur several times before the last larva pupates.

Photo by Johannes F. Skaftason

Order: *Diptera*

Horseflies

Tabanidae

Order: Diptera – Flies, suborder Brachycera; Family: Tabanidae – Horseflies and Deer Flies

Members of the Tabanidae family are commonly known as horseflies and deer flies in English. In Inuktitut, horseflies are known generally as *milugiaq* in Iqaluit and Igloolik. They are known as *kikturiaqsiuyuyuq* in Baker Lake.

Although this family includes both horseflies and deer flies, there are no reported sightings of deer flies in Nunavut, so only horseflies will be discussed in this section.

Description

Adults:
Horseflies are stout-bodied and have broad heads. They can be between nine and twenty-eight millimetres long. They have bulging, **iridescent** (with rainbow-like colours that shift and change in the light) **compound eyes**. Males and females can be distinguished by looking at their eyes. The male's eyes meet at the inner edges, while the female's eyes are separated. These flies have short antennae with only three

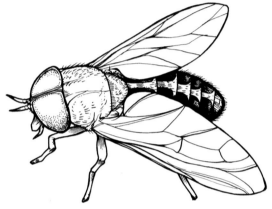

segments. The females have scissor-like **mandibles** for feeding on blood, while the males' jaws are adapted for sucking nectar. The bodies of both sexes are covered with dense, short hairs. Unlike some other flies that "buzz," horseflies are silent fliers.

Larvae:
The larvae are **fusiform**, meaning that they are broad in the middle but taper at each end (like fish and whales). They have eleven body segments. Their head capsules are retractable, which means that their heads can be pulled back into their bodies. They have antennae with three segments on their head capsules. The larvae are usually white, but they can also be brown or green. They may also have abdominal **prolegs**.

Life Cycle

Complete Metamorphosis:
The tabanid female usually lays eggs on a branch or on leaves that overhang wet ground or water. They lay their eggs several layers thick and can lay between one hundred and one thousand eggs over the course of a lifetime. The larvae hatch from the eggs in about one week and then fall down onto the ground or into the water.

Photo by Henri Goulet

Larvae go through six to nine **instars**. The larval stage can last from six months to a year. Tabanids usually **overwinter** as larvae in the soil. When the weather warms up the following spring, the larvae move up to the top part of the soil to **pupate**. The **pupal stage** generally lasts from two to three weeks. Males emerge before the females. After both sexes have emerged from the pupal stage as adult flies, they are ready to mate. Male tabanids chase the females. Mating starts in the air but is finished on the ground. The adult lifespan is thirty to sixty days in warmer climates. In Nunavut, this detail has not yet been studied.

Food and Feeding

Adult males feed on pollen and nectar. The females consume these plant products as well, but they also need blood meals, which they get by biting some types of mammals, including humans. The blood is needed to help with egg development. The larvae are **carnivorous**, feeding on other insect larvae, small crustaceans, and earthworms, if available, although there are no earthworms in Nunavut.

Habitat

The larvae do very well in damp situations—in swampy areas and by rivers, lakes, or even seashores. Not surprisingly, this is also where you will find most of the adults.

Range

These flies have been identified in Igloolik, Iqaluit, and Baker Lake by Inuit elders. They are most likely to be found in the more southern regions of Nunavut.

Did You Know?

When horseflies are plentiful, blood loss can be a real problem for some animals, such as moose, deer, horses, and cows. Some animals have lost up to three hundred millilitres of blood in a single day. Such blood loss can make the animals very weak.

Traditional Knowledge

Several elders from Baker Lake have indicated that horseflies seem to chase mosquitoes.

Balloon Flies

Empididae

Order: Diptera – Flies, suborder Brachycera

Family: Empididae – Balloon Flies, Dance Flies

Members of the Empididae family are commonly known as balloon flies or dance flies in English. No specific Inuktitut names were found for this family during the research for this book. Please see the Diptera account for general names for flies. Also, if you know of names for any species in this family, please contact the Nunavut Teaching and Learning Centre.

Description

Adults:
Balloon flies are small and bristly. They are between 1.5 and twelve millimetres long, and have round heads with large eyes and antennae with only three segments. They have large, stout, humpback **thoraces**, distinct necks, and long legs. They also have long, tapering **abdomens**, though the males' **genitalia** (sexual organs) are not folded forward under their abdomens, as they are in some other species of flies. These flies are mostly non-metallic.

Larvae:
Balloon fly larvae have visible heads with small antennae, as well as **palps** and **prolegs**.

Life Cycle

Complete Metamorphosis:
Not a lot is known about the habits of balloon flies. Most females deposit eggs, but some species are known to deposit larvae in dung. Once the adults emerge from the **pupae**, it is time for the flies to mate. Balloon flies are best known for their mating rituals. The males of many species capture smaller insects (mostly flies) and wrap them in silky, balloon-like structures. They carry these gifts into the swarm to attract females. This may also be a way to distract the females so they do not attack their mates. This type of gift-giving behaviour has been reported in the High Arctic.

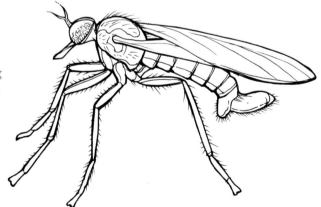

Food and Feeding

Although not a lot is known about balloon fly larvae, it is thought that they are **predaceous**. The adults may feed on flowers sometimes, but they mostly feed on other, smaller flies. It is thought that the adult female does not eat any food except for the gifts offered to her by the male during mating rituals.

Habitat

The larvae live in moist environments, such as under piles of dead leaves, in the soil, and on the surface of water. The adults tend to stay around the same sites, since this is where the females lay their eggs.

Range

Nineteen species have been reported on the tundra. At least five of these species live in the High Arctic.

Did You Know?

Many balloon flies have been found in pieces of amber (a yellow fossil resin from plant material). This means that these flies have been around since the Cretaceous period (from 65–140 million years ago).

Photo by Johannes F. Skaftason

Long-legged Flies

Dolichopodidae

Order: Diptera – Flies, suborder Brachycera; Family: Dolichopodidae – Long-legged Flies

Members of the Dolichopodidae family are commonly known as long-legged flies in English. No specific Inuktitut names were found for this family during the research for this book. Please see the Diptera account for general names for flies. Also, if you know of names for any species in this family, please contact the Nunavut Teaching and Learning Centre.

Description

Adults:
Long-legged flies are small to medium in size. They are usually metallic green, bluish, or copper in colour. These slender flies have bristle-like, three-segmented antennae, very large eyes, and short **proboscises**. The male's **genitalia** is often large and folded forward under his **abdomen**. As the name indicates, long-legged flies have long legs. The male's legs are longer than the females and are often ornamented with hairs or bristles. Long-legged flies have wings the length of their bodies that are usually clear, although sometimes they can have white or brown spots.

Larvae:
The larvae are **maggots**. They are slender, cylindrical, and white. Their front ends are tapered, while their back ends are cut short. The larvae have bumps on segments four to eleven. Their unhardened heads have both antennae and **palps**.

Pupae:
Larvae construct their **pupae** from bits of debris. There are two long respiratory horns at the top of the **thorax** through which the insect breathes as it changes to its adult form.

Life Cycle

Complete Metamorphosis:
Long-legged flies are very sensitive to cold weather and therefore only come out in the late spring or early summer. After the white-to-brown, oval-shaped eggs are laid and then

hatch, the larvae develop through several **instars** in wet to dry soil. The larvae **pupate** in **cocoons** made from soil particles cemented together. After their emergence from the pupae, the adults begin their mating rituals. Some males have elaborate mating behaviour. Depending on the species, they may have some type of decoration on their legs, which they wave around to attract females. They may have fans or disks of white or black scales on their legs. Some males also have wings patterned in a way that attracts females. Other males have territories, which they defend from intruders. Some males even hold on to the females after they have mated so that the females cannot mate with a second male.

Food and Feeding

Photo by Johannes F. Skaftason

The majority of adult long-legged flies **predaceous**, preying in large part on soft-bodied insects (adults or larvae). Most of the adults grab their prey with their front legs and start to chew. They then suck up the liquids that this chewing creates and leave what cannot be sucked up. Not much is known about the feeding habits of larvae. However, it is thought that they also catch and eat other insects or **scavenge**, eating things that they find already dead.

Habitat

Most long-legged flies are associated with wet places; for example, stream and lake margins, salt marshes, seashores, and humid forests (of course not in Nunavut, where there are few forests). Larvae are found in mud, damp soil, decomposing leaves, moss, seaweed, and algae mats, and under tree bark (although, once again, not in Nunavut).

Range

Twenty-eight species have been reported in Arctic North America. Approximately eight of these species are found in treed areas to the west of Nunavut. Two species have been collected in the High Arctic.

Did You Know?

Some recent studies have demonstrated that long-legged flies have very specific habitat requirements. Populations of these flies react quickly to small perturbations in their habitats, making them possible bio-indicators of deteriorating or changing environments.

Hoverflies

Syrphidae

Order: Diptera – Flies, suborder Brachycera; Family: Syrphidae – Hoverflies

Members of the Syrphidae family are commonly known as hoverflies in English. No specific Inuktitut names were found for this family during the research for this book. Please see the Diptera account for general names for flies. Also, if you know of names for any species in this family, please contact the Nunavut Teaching and Learning Centre.

Description

Adults:
Hoverflies come in a large variety of shapes, colours, and sizes. The flies are between four and twenty-five millimetres long. Some are bright, with yellow, brown, and black stripes, while others are simply black or brown. Some hoverflies are smooth and hairless like wasps, while others are hairy like bees. In fact, you might have to look twice at these flies to make sure they are not bees or wasps. Their large eyes seem to cover their entire heads. They have short, fleshy **proboscises** and short antennae. Hoverflies also have folds or false veins in the middle of their wings that help entomologists to identify them. They are really good fliers and are able to hover.

Larvae:
Hoverflies produce three types of larvae, depending on the species. The aphid-feeding larvae are **maggot**-like but greenish in colour. Larvae in polluted **aquatic** environments have elongated hind ends that look like tails. These tails are actually breathing tubes and these larvae are referred to as rat-tailed maggots. The larvae that live in ants' nests, none of which occur in Nunavut, are oval-shaped and flat.

Life Cycle

Complete Metamorphosis:
Hoverflies lay their single eggs close to sources of food for the larvae, such as aphids, plant bulbs, dung, or even decaying flesh. They may lay up to several hundred eggs during the

summer. In two to three days, once the eggs hatch, there is plenty of food for the small, legless maggots. They feed and go through two **instars** before they finally **pupate**. This takes about two to three weeks. Hoverflies overwinter as **pupae**. The following spring, the adults emerge from the pupae and the cycle begins again.

Food and Feeding

The adults feed mainly on nectar and pollen. Some larvae eat decaying materials in ponds, streams, or soil. Some eat aphids, dung, or fungi, while others are filter feeders in the water. Some even act as **parasites** in the nests of ants. In the Low Arctic, hoverfly larvae feed mainly on aphids, while in the High Arctic, they feed on both decaying materials and aphids.

Photo by Johannes F. Skaftason

Habitat

Adult flies are often seen hovering, or feeding on flowers on the tundra. The larvae are found either on the branches or leaves of plants where aphids occur, or living in decaying organic matter.

Range

Five species of syrphid flies have been reported in the High Arctic and at least another dozen in the Low Arctic.

Did You Know?

Hoverflies look a lot like bees and wasps, and they can hover like them as well. This is called mimicry. Hoverflies pretend they are bees and wasps to warn potential **predators** to avoid eating them.

Photo by Johannes F. Skaftason

Order: *Diptera*

Dung Flies

Scathophagidae

Order: Diptera – Flies, suborder Brachycera; Family: Scathophagidae – Dung Flies

Members of the Scathophagidae family are commonly called dung flies in English. Along with most flies, dung flies can be called *anangiq* or *niviuvak* in Inuktitut. In Baker Lake, dung flies are also called *kikturiaqsiuti*. In Kugluktuk, dung flies are *milogiakyoak* or *milogiak*, while in Iqaluit, they are known as *qaumaja*. In Taloyoak, an elder identified the dung fly as *niaqusiugjujuq*.

Description

Adults:

Dung flies are between three and eleven millimetres long and can be black, grey, brown, or yellow. They have round heads and widely separated **compound eyes**. Their antennae are short, having only three segments.

Dung flies, depending on the species, have either piercing or non-piercing mouthparts. They also have well-developed **palps**. These flies have slender bodies and legs. Their bodies can be strongly to weakly bristled and their legs are hairy. The wings of dung flies are usually clear, but some species do have spots or cloudy sections. Their lower **calypters** are either reduced or completely absent.

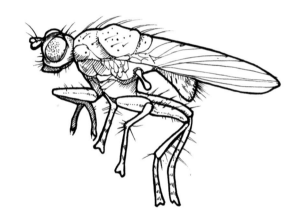

Larvae:

The larvae of dung flies are almost cylindrical, usually tapering at their front ends. Their **mandibles** can be thin or thick, and their **abdomens** have minute spines. Their hind ends have four to eight pairs of **tubercles** (projections or outgrowths) varying in size and arrangement, as well as **spiracles** (breathing spores). The breathing spore at the hind end is attached to a short tube.

Life Cycle

Complete Metamorphosis:

Depending on the species of dung fly, the female will either attach the pale, oval-shaped egg to the surface of a leaf, or insert the egg into plant tissues, faeces, or other available materials. Once the larvae hatch, they feed and go through five to eight **instars**. The **pupa** forms inside a **puparium**, a case that is created by the hardening of the final larval skin. The adult emerges from the pupa and the cycle starts again.

Food and Feeding

Adult dung flies feed on insects or other **invertebrates**. Some dung fly larvae eat plants and are particularly fond of sedges (*Scirpus*), rushes (*Juncus*), and louseworts (*Pedicularis*). Other larvae capture and eat smaller insects. Those of the genus *Scathophaga* feed on dung or decaying seaweed.

Habitat

The larvae can live in a variety of habitats, depending on what they eat. They can live by the sea; in water, wet soil, or rodent burrows; or on their preferred plants. The adult flies live in the same areas since those are the places where they lay their eggs.

Photo by Johannes F. Skaftason

Range

There are about 150 flies from this family recorded in Canada. Of those, approximately twenty-five species are mostly restricted to the Arctic. At least five species have been found in the High Arctic.

Did You Know?

Photo by Johannes F. Skaftason

Even though the flies in this family are commonly referred to as dung flies, there are only a few species that actually feed on dung and spend their larval stage there (these few are from the genus *Scathophaga*).

Photo by Johannes F. Skaftason

Order: *Diptera*

Root-maggot Flies

Anthomyiidae

Order: Diptera – Flies, suborder Brachycera; Family: Anthomyiidae – Root-maggot Flies

Members of the Anthomyiidae family are commonly known as root-maggot flies in English. No specific Inuktitut names were found for this family during the research for this book. Please see the Diptera account for general names for flies. Also, if you know of any names for species in this family, please contact the Nunavut Teaching and Learning Centre.

Description

Adults:
Root-maggot flies are between two and twelve millimetres long. They have antennae with two to six segments, above which their eyes meet. Their mouthparts are functional and each fly has one segmented **palp**. These flies can be yellow, grey, brown, or black, and are non-metallic in colour. Their **abdomens** are cylindrical and have four to five visible segments. Root-maggot flies have hairy, black-to-yellowish legs. The male's legs are hairier and longer than the female's. Their wings are sometimes clouded with grey or brown and they have well-developed **calypters**. One of the features distinguishing this fly from the housefly (they are otherwise very similar) is that its 2A vein, the last vein on the bottom of the wing, extends all the way to the edge of the wing.

Larvae:
The larvae are slender, smooth **maggots**. They have no visible heads or legs. They are blunt at one end and pointed at the other. The blunt end has tiny spines in a circle.

Life Cycle

Complete Metamorphosis:
Once the eggs are deposited and the larvae hatch, they go through three **instars**, and then feed for several weeks before they **pupate**. The pupae **overwinter**, and in the spring the flies emerge and the cycle starts again.

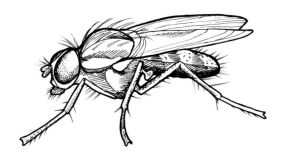

Food and Feeding

The adults feed on nectar from flowers. The larvae of many species in the South feed on the roots of vegetables, making them serious pests. It is more likely for Arctic species to feed on decaying organic matter, such as seaweed or other decaying materials further inland.

Habitat

The larvae live in damp, even wet, habitats inland or in rotting vegetation on the seashore. The adults live nearby, but they also need flowers for food, which are not necessarily found on the seashore or in wet habitats. Instead, you will see them around flowers on the tundra.

Range

About 130 species of root-maggot flies have been identified in the North American Arctic. It is hard to estimate exactly how many of those species are found in Nunavut. About seventy occur in the Eastern Arctic, with at least six being recorded in the High Arctic.

Did You Know?

Root-maggot flies look a lot like houseflies, and at one time they were put in the same family, Muscidae. Now, thanks to better equipment, entomologists that study flies have put them in two separate families.

Photo by Andre Karwath

Order: *Diptera*

Houseflies

Muscidae

Order: Diptera – Flies, suborder Brachycera; Family: Muscidae – Houseflies

Members of the Muscidae family are commonly known as houseflies in English. In Iqaluit, you might find houseflies referred to as *milugialaaq, anangiqpa*, or *anangirjuaq*. In Kugluktuk, they have different names, such as *milookatak* or *niviovak*. Similarly, in Taloyoak and Baker Lake, they are called *niviuvak*.

Description

Adults:
Houseflies are stocky and heavily bristled, and measure between three and twelve millimetres long. They have feathery antennae that are shorter than their faces. Some houseflies have sponge-like mouthparts, while others have lance-like projections for biting. They have large **compound eyes**. Houseflies are usually dull in colour—either black, grey, or yellowish—but there are a few species that are metallic blue or green. They have slender legs and wings without any markings. Houseflies taste things with their feet.

Larvae:
The larva is almost cylindrical, with a tapered front end. The hind end is blunt. It does not have a distinct head, although it does have **mandibles** pressed close to the body. The larva's **abdominal** segments have welts, and sometimes it has pairs of **prolegs** on segment eight of its **abdomen**.

Life Cycle

Complete Metamorphosis:
Female houseflies lay their eggs where the larvae will, once hatched, find food easily. This could be on dung, decaying vegetation, meat, or elsewhere. Each female can lay up to five hundred eggs in small batches. The larvae usually go through three **instars** as they grow and develop. The **pupae** are enclosed in a **puparium**, which is made from the final instar skin. This is where the larvae transform into adults. Emerging adults live for fifteen to twenty-five days.

Photo by Johannes F. Skaftason

Food and Feeding

Houseflies feed on dung, pollen or nectar from flowers, and dead or decaying plant and animal matter, and some houseflies hunt and eat other insects. Some of the flies in this group also feed on blood and any other kind of fluid or liquid that oozes from a decaying corpse.

Habitat

The larvae live in many different habitats, depending on diet. They can be found in dung or decaying vegetation—anything from fruit to logs, fungi, fresh water, and **carrion**. The adults are usually found in the same kinds of environments, since they need to find food for their larvae as well as for themselves. Houseflies are also often found around human habitation, where there is plenty of food for them.

Range

Over 160 species of houseflies have been reported from Arctic North America. About twenty of these were found in the High Arctic.

Photo by James Lindsey

Did You Know?

Because they breed in decaying flesh or dung, flies from this family are often associated with diseases. Flies can be vectors of certain diseases, meaning that they can act as carriers of disease-causing micro-organisms from one host to another. For example, in mid-continental Africa, the tsetse fly transmits a protozoan **parasite** that causes what is known as sleeping sickness.

Order: Diptera

Blowflies

Calliphoridae

Order: Diptera – Flies, suborder Brachycera; Family: Calliphoridae – Blowflies

Members of the Calliphoridae family are commonly known as blowflies in English. No specific Inuktitut names were found for this family during the research for this book. Please see the Diptera account for general names for flies. Also, if you know of names for any species in this family, please contact the Nunavut Teaching and Learning Centre.

There are only seventy-five species of Calliphoridae known in North America.

Description

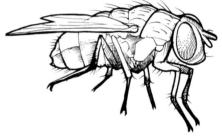

Adults:
Blowflies are small to medium in size, ranging from five to fifteen millimetres long, and are metallic green or blue. They are robust flies with heads wider than they are high. They have antennae with three segments and **aristas** (thread-like attachments) at the tips. Some males have eyes that meet above the antennae, while others do not. These flies have short **proboscises**. The blowfly has thick legs with strong bristles on its hind **tibiae** (the fourth segments of its legs). It also has bristles on both sides of the **thorax** and on the tip of the **abdomen**. It has a well-developed lower **calypter**.

Larvae:

There are two types of Calliphoridae **maggots**: hairy and smooth. Hairy maggots are not truly hairy, but they have **papillae** (little protrusions on the body) that make them look hairy. Both hairy and smooth maggots have well-developed mouth hooks. The larvae are pale yellow to white and cylindrical, and have tapered front ends. They have no real heads.

Life Cycle

Complete Metamorphosis:
The blowflies that humans are most familiar with, the bluebottles or greenbottles, typically lay their eggs indoors on fresh or cooked meat, fish, or even dairy products. They also lay eggs in the open on any kind of carcass they may find. The egg clusters look like miniature rice balls. One of the blowflies from the genus *Calliphora* (*Apaulina sapphira*), found in the High Arctic, lays its eggs in birds' nests and, when the larvae are born, they suck blood from the chicks.

The process of hatching from eggs laid in carcasses to becoming larvae only takes about one day. The larvae go through three **instars** as they grow. When the third instar has finished growing, it leaves the corpse it has been feeding on and falls to the ground. It burrows into the soil, forms a **pupa**, and, seven to fourteen days later, the adult fly emerges. At least some of these flies **overwinter** as adults.

Food and Feeding

The adults feed on nectar, **honeydew** (produced by aphids), and the liquids that are produced when something organic or living decomposes. Blowflies can pick up odours of decay from quite far away. They are able to fly up to twenty kilometres in search of a corpse. The adults prefer flowers with strong odours—there are actually some flowers that smell like rotting meat. The larvae of most species feed on **carrion** and dung.

Photo by Johannes F. Skaftason

Habitat

Adult flies tend to hang around human habitations, since food is easily accessible there. The larvae live mostly on corpses and dung.

Range

Photo by Johannes F. Skaftason

About twelve species have been reported from the tundra, with three of these living in the High Arctic. One High Arctic species, *Protophormia terranovae*, can cause **myiasis** (any disease resulting from a maggot infestation of the tissues or cavities of the body) among caribou. As mentioned before, *Apaulina sapphira*, another High Arctic species, has parasitic larvae that live on nestlings. The third High Arctic species, *Boreallus atricups*, lives on carrion, as do all of the Low Arctic species.

Did You Know?

The development of blowflies is very predictable if the surrounding temperature is known. Because of this, they are a valuable tool for forensic scientists in determining the time of death of a corpse that blowflies have invaded.

Traditional Knowledge

From Kugluktuk: Annie Kellogok interviewing Joseph Niptanatiak, July 2006
The niviovaks are good for the land: when there are dead animals they eat them up and clean the land. But people do not like them because they lay eggs on our food and spoil the food. They are good for our environment though.

Warble Flies

Oestridae

Order: Diptera – Flies, suborder Brachycera; Family: Oestridae – Warble Flies, Botflies

Members of the Oestridae family are commonly known as warble flies and botflies in English. In Inuktitut, in all the communities consulted, *iguttaq* (*igutsaq, egotak*) is used to name both warble and botflies. This is the same name used for bees. In Iqaluit, warble flies and botflies were also referred to as *milugiaq* and *arlungajuq*.

There are two species of Oestridae found commonly in the Canadian Arctic, and they are significant because they **parasitize** caribou. One is a warble fly that lives under the skin at the hind end of the caribou, while the other is a botfly that makes its home in the nasal cavities of the caribou. They will be treated together in this account because they are very similar except for their chosen location on the caribou's body. This difference will be pointed out in the life cycle section.

Description

Photo by P.G. Penketh

Adults:
Warble and botflies are large, measuring between 0.9 and twenty-five millimetres. They are stout-bodied and resemble bees. They are hairy—even their **thoraces** may have long hair—and they have broad heads and flat faces with small eyes. Their antennae are short and they have small or atrophied (poorly developed) mouthparts. Warble and botflies have round **abdomens** and stubby, hairy legs.

Larvae:
The larvae are white to yellowish, and do not have heads. Their **mandibles** are well-developed. The larvae are elongated and flat on their undersides, with spines on their backs.

Life Cycle

Complete Metamorphosis:
Hypoderma tarandi: Caribou warble fly
In midsummer, warble flies lay their eggs on the hair of caribou legs or on their lower bodies. Once the eggs hatch, the larvae make holes in the skin and live just under the skin. They slowly make their way to the caribou's back during the fall and winter. Once they reach their destination, the larvae cut breathing holes through the skin. Here, they go through three **instars**. In early summer, the larvae are ready to **pupate**, which they fall to the ground to do. The adult flies emerge from the **pupae** and only live long enough (three to five days) to mate and lay eggs.

Cephenemyia trompe: Nasal botfly
The females deposit larvae (not eggs) around the nostrils of the **host** caribou. They then crawl into its mouth and move to its nasal cavities or the back of its throat, where they spend the winter. The following spring, the larvae are ready to pupate and, as the caribou sneezes and snorts, the larvae exit the body. The larvae pupate on the ground, and the adults emerge shortly afterwards.

Photo by Henri Goulet

Food and Feeding

The adults do not feed but rather survive on fat stored in their bodies. They only live a few days. The developing larvae feed on blood and other liquids in the caribou's body.

Habitat

In the North, adult flies can be found flying around caribou, while larvae live under their skin or in their noses and throats. In the South, different species in this family parasitize animals such as sheep, horses, and cows.

Range

The warble and botflies of Nunavut attack caribou in the Low Arctic. Warble flies have also been found in the High Arctic on Peary caribou.

Did You Know?

When a warble fly makes a breathing hole in the caribou's skin, this creates a cyst (thin-walled, hollow cavity) that the larva lives in until it is time to pupate.

Traditional Knowledge

After hunting caribou, while butchering the meat, Inuit might have, at one time, consumed the larvae found under the caribou's skin. The preference for eating them once varied regionally, but it seems that this is no longer practised much anywhere.

Tachinid Flies

Tachinidae

Order: Diptera – Flies, suborder Brachycera; Family: Tachinidae – Tachinid Flies

Members of the Tachinidae family are commonly known as tachinid flies in English. No specific Inuktitut names were found for this family during the research for this book. Please see the Diptera account for general names for flies. Also, if you know of names for any species in this family, please contact the Nunavut Teaching and Learning Centre.

Description

Adults:
Tachinid flies are very diverse in appearance. They are between two and twenty millimetres long, with lots of **bristles**. They have three-segmented antennae with bare **aristas**, as well as **compound eyes**. Their mouthparts are well-developed and functional. These flies can be grey or black, or they can be brightly coloured like wasps. One of the identifying features of this kind of fly is its well-developed **subscutellum**

(ridge or lobe under the **scutellum**, which is the hardened thorax). Both the **scutum** (hard part of the thorax in front of the scutellum) and the **abdomen** of the tachinid fly are covered with bristles. It also has a large **calypter**.

Larvae:
Tachinid larvae are **maggot**-like. The first **instar** has a hook with which it holds on to the host and begins the burrowing process. The second and third instars have **mandibles** with which they eat. Some larvae are very flat and have sticky substances on their abdomens.

Life Cycle

Complete Metamorphosis:
All known larvae of the tachinid family are **parasitoids** (organisms that live and feed off host organisms until the hosts are eventually killed) of insects and other **arthropods**. Tachinid flies have evolved different techniques for parasitizing the **hosts** that they attack.

The most primitive of these flies attach eggs to host insects. The eggs have sticky undersides that help them stay attached. During the time it takes for the eggs to hatch, they are vulnerable to being eaten by their hosts or even being knocked off. Once the eggs hatch, the larvae burrow into the hosts and continue their development inside, all the while gradually eating and killing the hosts. Only a small percentage of tachinid flies use this method.

Another group of tachinid flies have evolved so that they can easily insert their eggs partially or completely into the bodies of hosts, thus ensuring that the eggs will not get eaten or knocked off. Development of these larvae inside their hosts continues as described above.

Photo by Ron Hemberger

Because some of the hosts have developed spines, hairs, and webbing, and have even become **nocturnal** to protect themselves from such attacks, tachinid flies have had to adapt as well. This is similar to an arms race between opposing forces, each one trying to develop better machinery or techniques. In this race, many tachinid flies have developed **ovisacs**, sacs in which the eggs are kept until the larvae are almost ready to come out. The eggs hatch within seconds of being laid on the host and begin to burrow right away. Some members of the tachinid family employ still another means completely. They locate damaged leaves where a possible host has been feeding, and lay their eggs nearby. After the larvae hatch, they wait to ambush that possible host. So that they don't dry out while they are waiting, these larvae have series of **sclerites** (hardened plates) that fit like suits of armour, keeping the moisture in.

Food and Feeding

Adult tachinid flies feed on nectar in flowers and on **honeydew**, whereas larvae are parasitoids of other insects. In the Arctic, they mostly feed on the larvae of butterflies and moths.

Habitat

Adults are commonly seen flying around flowers when they are feeding and around other plants when they are looking for places to lay their eggs. The larvae spend all of their time inside the host until they **pupate** and the adult flies emerge.

Range

At least seven species of tachinid flies have been identified in the Arctic. Most of these occur in both the High and Low Arctic.

Did You Know?

Parasitoids live and feed off other animals until those animals die, while parasites feed on other animals, but do not kill them. Most tachinid flies are parasitoids.

Butterflies and Moths

Order: Lepidoptera

There are two hundred thousand species of butterflies and moths worldwide. The name Lepidoptera comes from the Greek words *lepis*, meaning scales, and *ptera*, meaning wings.

The general names for butterflies and moths in Baker Lake, Taloyoak, Iqaluit, Kugluktuk, Clyde River, and Igloolik are very similar: *haqalikitaaq, saqakilitaaq, taralikitaaq, hakalikitaak, tarralikisaaq,* and *taqralikisaq*.

Caterpillars (the larvae of lepidopterans) are called *qulluriaq* or *miquligia* in Iqaluit. In Kugluktuk, they are called *koglogiak*. In Taloyoak, the word for caterpillar is *autviq*. In Baker Lake, a caterpillar without hair is called *quglugiaq*, while one with hair is called *aahjiq*. In Clyde River, *miqquligiaq* is the common name for a caterpillar.

Description

Adults:

Lepidopterans come in a wide variety of shapes and colours and can have wingspans anywhere from three millimetres to thirty-two centimetres. Each has a **haustellum**, which is a long, tongue-like sucker. The haustellum is rolled up under the head when the insect is not feeding and is extended when it is. Some are longer than others, depending on the type of flowers at which the insects' feed.

Photo by Thomas Bentley

Lepidopterans have large **compound eyes** with vision that allows them to see in colour. Their antennae are long and slender and function both as noses and as taste buds. Their **palps** are well-developed. Many females have two genital openings at the ends of their **abdomens**. In one opening, the male deposits the sperm; from the other, the female lays the eggs.

Lepidopterans have three pairs of legs and two pairs of wings, all attached to the **thorax**. The ends of the legs are capable of detecting moisture, sugar, and even certain chemicals. The hindwings are a little smaller than the front wings. The wings are **membranous**, but they are covered in **scales**, which appear like a fine dust on your fingers if you touch them. The head and legs are covered with scales or modified scales as well. If the wings were not covered with scales, they would be transparent, like the wings of other insects. The scales are hollow except in the most primitive species. The veins of the wings are pumped full of air and fluids to help maintain the structure and shape of the wings when flying. Lepidopterans have big flight muscles.

Larvae:
Caterpillars have soft bodies and well-developed heads. On their heads, they have very short antennae and six **ocelli** on either side. They have biting or chewing mouthparts, with silk glands and **spinnerets** (organs through which silk is produced) on either side. Their bodies are composed of thirteen segments—three making up the thorax and ten making up the abdomen. Caterpillars' bodies are often covered with hair-like or other projections. They have three pairs of true legs and up to five pairs of fleshy **prolegs**. These prolegs have tiny hooks at the tips.

Life Cycle

Complete Metamorphosis:
Courtship takes place in the air as the male follows the female or flies above her. If she is interested, she lands and mating takes place. Next, the female searches for an appropriate place to lay her eggs. She needs to find a place where plenty of food will be available to her babies. The females can lay eggs singly or in groups on the undersides of leaves or on stems.

The larvae, or caterpillars, hatch and usually go through five **instars** before they **pupate**. The caterpillar usually lives for three to six weeks while it eats and eats. After its final **moult**, the caterpillar spins a **cocoon** if it is a moth. If it is a butterfly, it forms a **chrysalis** without a cocoon. Some lepidopterans **overwinter** as pupae, while others develop into butterflies or moths after ten days.

Food and Feeding

Some adults feed on nectar in flowers, while a few do not eat at all. Most larvae feed on plants. However, there are a few species that eat other insects or dead plant materials.

Habitat

Since butterflies and moths feed on nectar, they are most commonly found flying near, or resting on, flowering plants on the tundra. Caterpillars are also mostly found on the plants that they eat.

Range

There are approximately thirty-two species of tundra-adapted butterflies. Seven of these survive on the High Arctic islands. Just over one hundred species of moths live in the Arctic, at least twelve of which can be found in the High Arctic.

Order: Lepidoptera

Moths	Butterflies
Fly at night	Fly during the day
Do not have clubbed antennae	Have clubbed antennae
Are dull-coloured	Are brightly coloured
Hold wings flat across their backs when at rest	Clap wings upright when at rest
Have tiny hooks to link front wings to hindwings	Employ a different system to keep wings together in flight

Did You Know?

There are many more species of moths than there are butterflies. Although moths and butterflies belong to the same order, there are differences that generally distinguish these two types of lepidopterans.

Traditional Knowledge

From Baker Lake: Brenda Qiyuk interviewing Silas Aittauq, July 2006

I [did not do] this myself, but I remember how children used to chase moths and butterflies and try to give them to the Catholic priest in exchange for candy. When children brought live moths or butterflies to the Catholic priest, he would exchange them for candy, so whenever children wanted candy, they would try to catch butterflies. The Catholic priest wanted them for medicine. I don't know exactly what kind of medicine. All I heard was that moths and butterflies were needed for medicinal use.
Several of the elders in Baker Lake had this same story.

Moth resting

Butterfly resting

Order: *Lepidoptera*

Leafroller Moths

Tortricidae

Order: Lepidoptera – Butterflies and Moths; Family: Tortricidae – Leaf Roller Moths

Members of the Tortricidae family are commonly known as leafroller moths in English. No specific Inuktitut names were found for this family during the research for this book. Please see the Lepidoptera account for general names for butterflies and moths. Also, if you know of names for any species in this family, please contact the Nunavut Teaching and Learning Centre.

Description

Adults:
Leafroller moths are stout-bodied, with a wingspan between eight and forty millimetres. They have well-developed **proboscises** and threadlike antennae. These moths usually have **ocelli**, as well as **compound eyes**. The front wings are square at the tips, and the wings can be mottled or have dark bands. Leafroller moths can be grey, tan, or brown.

Larvae:
The larvae, commonly known as caterpillars, are small and smooth-skinned, with well-developed, **sclerotized** heads. They are slender and soft-bodied, and have thirteen segments. They vary in colour from green to brown, with darker coloured heads.

Life Cycle

Complete Metamorphosis:
The female leafroller moth lays her eggs on a leaf. Once the eggs hatch, the small caterpillars eat and eat. Some of these caterpillars roll themselves inside leaves and eat in the safety of these hiding places. They go through five to six **moults** before they **pupate**. Again, some pupate inside rolled leaves for protection. The leaves are held in the rolled position by spun pieces of silk. Some leafroller moth species **overwinter** as pupae, while others overwinter as eggs. The adult moths emerge from the pupae.

Food and Feeding

Caterpillars of two species have been found feeding on louseworts in the High Arctic, as well as on mountain avens. The adults do not feed.

Habitat

Leafroller moths and their larvae can be found near their favourite food sources on the tundra.

Range

Although fourteen species have been reported in Arctic North America, six of those live in forested areas and therefore not in Nunavut. Three species have been recorded in the High Arctic.

Did You Know?

When a leafroller caterpillar is disturbed, it will often drop a silken thread and bungee jump to another location.

Traditional Knowledge

From Taloyoak: Ellen Ittunga interviewing Mary Ittunga, July 2006

> *Butterflies were also used on newborns. Inuit believed that, if you placed or rubbed a butterfly on a newborn girl, she [would] grow up to make beautiful designs when she sew[ed].*

A similar story is reported from Baker Lake, except this time using a caterpillar and not a butterfly or moth.

Brenda Qiyuk interviewing Silas Aittauq, July 2006

> *There is this one bug that was used to help a girl become good at sewing. They would kill the quglugiaq [caterpillar] and spread the skin on the young girl's arm so she would be talented at sewing clothing.*

Photo by Tom Murray

Order: Lepidoptera

White and Sulphur Butterflies
Pieridae

Family: Pieridae – White and Sulphur Butterflies

Members of the Pieridae family are commonly known as white and sulphur butterflies in English. Only one specific Inuktitut name was found for this family during the research for this book. Some of the elders mentioned that a yellow butterfly could be prefaced by *quqsuqtuq*, which means yellow in Inuktitut. The whole name for this butterfly in Iqaluit would then be *quqsuqtuq tarralikitaaq*. Please see the Lepidoptera account for general names for butterflies and moths. If you know of additional specific names for members of this family, please contact the Nunavut Teaching and Learning Centre.

Description

Adults:

White and sulphur butterflies have a wingspan between twenty-two and seventy millimetres. They have distinctly clubbed antennae. These butterflies are white, yellow, or orange, with rounded wings and simple patterns. For instance, they might have one black spot on one wing or a black border around both wings. Males and females of this family have different colouring. The front legs are well-developed, and are used for walking. The claws at the end of the leg are forked.

Larvae:

The caterpillars are soft-bodied, with no hairs or spines. Each has a distinct head, three pairs of true legs, and up to five pairs of **prolegs**. They are generally green or reddish-brown.

Life Cycle

Complete Metamorphosis:

Female white and sulphur butterflies lay eggs on leaves, stems, or buds. The eggs are elongated and ribbed. Out of the eggs hatch the larvae, which are commonly known as caterpillars. The caterpillars eat and eat, **moulting** four times as they grow. At the end of the final **instar**, they change into **pupae**, or **chrysalids** as butterfly pupae are known. The pupae are attached to plants with silk. Many Pieridae butterflies **overwinter** as chrysalids, and the cycle begins again in the spring with the emergence of the adults.

Food and Feeding

The caterpillar larvae of the *Colias* species, one of the butterflies in this group, feed on yellow oxytrope, alpine milk-vetch, and blueberry, cranberry, and willow leaves. The adult butterflies feed on flower nectar.

Habitat

Like most butterflies and moths, members of this family are most frequently found near their favourite foods—in this case, tundra plants. Adults and larvae live in the same habitats.

Range

Fewer than ten species of Pieridae butterflies live as far north as the Eastern Arctic, and only two of those live in the High Arctic.

Photo by Mary Hopson

Did You Know?

Some of these butterflies exhibit sexual dimorphism, which means that the males and females are not identical. If we think of some birds, for instance, males are often beautiful and brightly coloured, while the females are drab or plain. In Pieridae butterflies, the differences are not as significant as in some birds, but they are still present.

Traditional Knowledge

From Clyde River: Rebecca Hainnu interviewing Peter Kunilusie, July 2006

Peter explains the different names for the different colours and patterns of butterflies and moths.

> *There are yellow ones, and ones with spots. The ones with spots are called* kakiattut. *The ones that don't have yellow, we call them* tuurngaviaq. *The ones that are yellow or blue are* tarralikisaaq.

Order: *Lepidoptera*

Gossamer-winged Butterflies

Lycaenidae

Family: Lycaenidae – Gossamer-winged Butterflies

Members of the Lycaenidae family are commonly known as gossamer-winged butterflies in English. No specific Inuktitut names were found for this family during the research for this book. Please see the Lepidoptera account for general names for butterflies and moths. Also, if you know of a specific name for any species in this family, please contact the Nunavut Teaching and Learning Centre.

Description

Adults:
Gossamer-winged butterflies are small and brightly coloured, and have a wingspan of eleven to forty-seven millimetres. Their antennae are usually ringed with white, and they have lines of white scales around their eyes as well. These butterflies have slender bodies. The male's front legs are reduced in size and are without claws. The female's legs are normal. Their wings are covered with pigmented, light-refracting **scales**, which give them a metallic sheen.

Two subfamilies of the Lycaenidae family make it to the Arctic: the coppers (subfamily Lycaneninae) and the blues (subfamily Polyommatinae). The coppers are reddish-brown with a coppery tinge. They have black markings and are very fast fliers. The blues' upper wing surfaces are completely or almost completely blue, the female's being darker than the male's. Many of these butterflies secrete a **honeydew** substance.

Larvae:
The larvae, commonly known as caterpillars, are slug-like but flat. They are small and hairy. They may also have glands that produce secretions which attract ants.

Life Cycle

Complete Metamorphosis:
The males actively fly around their favourite plant foods looking for receptive females. Once fertilized, the females lay eggs on or near the petals or the undersides of plant leaves, which provide food for the larvae once they hatch. Some species **overwinter** as eggs, while

others go through **diapause** (a long break during which the body functions slow down and the insect waits for the right conditions under which to continue its life cycle) as **pupae** or even larvae. Once the larvae reach the final stage of growth, they pupate. The pupa, or **chrysalis** as it is called for butterflies, is suspended by a silk **girdle** (a circular string that surrounds the chrysalis). The adults emerge from the chrysalids and the cycle begins again.

Photo by Thomas Bentley

Food and Feeding

The larvae of different species feed on different plants. Of the species found in Nunavut, the American copper larvae (*Lycaena phlaeas*) feed on mountain sorrel; the Cranberry blue larvae (*Vacciniina optilete*) feed on blueberry and cranberry plants; and the Arctic blue larvae (*Agriades glandon*) feed on *Diapensia lapponica* (pincushion plants), crowberry plants, and purple saxifrage. The adults all drink nectar from flowers.

Habitat

These butterflies are most often seen on the tundra near the plants they feed on as both larvae and adults.

Range

Only three species of these colourful butterflies have been reported in Nunavut. The American copper is found both on the mainland and the Arctic islands—all the way up to Ellesmere Island in the High Arctic. The cranberry blue can be spotted on the mainland, around Rankin Inlet and Arviat, and also in the Kugluktuk area. The Arctic blue can be found all around Nunavut, both in the High and Low Arctic.

Photo by Mary Hopson

Did You Know?

Many caterpillars in the Lycaenidae family have special relationships with ants. The caterpillars have glands which produce a sweet honeydew that ants collect. In exchange, ants protect these caterpillars from **predators**. Aphids and ants have a very similar relationship to one another. This exchange, however, does not happen in Nunavut, since ants do not make it that far north.

Brush-footed Butterflies
Nymphalidae

Order: Lepidoptera – Butterflies and Moths; Family: Nymphalidae – Brush-footed Butterflies

Members of the Nymphalidae family are commonly known as brush-footed butterflies in English. No specific Inuktitut names were found for this family during the research for this book. Please see the Lepidoptera account for general names for butterflies and moths. Also, if you know of specific names for any species in this family, please contact the Nunavut Teaching and Learning Centre.

Description

Adults:
Brush-footed butterflies can range from small to large in size, but most of them are medium-sized. These butterflies have two grooves on the undersides of their **clubbed** antennae. In North America, the most common colour of brush-footed butterflies is orange, but there are many exceptions and they can vary greatly in appearance. The undersides of the wings

Photo by Carolyn Mallory

are dull and resemble dried leaves, which is useful for camouflage when the butterflies are at rest. This is always the case in the males of the family, and there are only a few exceptions among females. They have long **proboscises** and reduced front legs. The front legs are covered in long hair, giving them the appearance of brushes. These legs are useless, and are held up near the face. They are too small to be used when walking, so these butterflies walk on their middle and back legs.

Photo by Mary Hopson

Larvae:
There is great variation among the larvae (also known as caterpillars) of the different subfamilies of Nymphalidae. However, all of the caterpillars are cylindrical and either hairy or spiky, and have projections on the head.

Life Cycle

Complete Metamorphosis:
In the Fritillary subfamily, the females lay eggs singly in the debris near food plants or on the undersides of their leaves and on their stems. Females in the Tortoiseshell subfamily lay their eggs in clumps on their preferred plants. Members of the Arctic subfamily and the Alpine subfamily lay eggs singly on or near grasses and sedges. The caterpillars emerge from the

eggs and begin to eat their favourite plants.

In many of the northern species, the caterpillars take more than one year to become butterflies. When this is the case, the early caterpillars hibernate until the following summer. Then they go through several **moults** and hibernate again for another winter. Once the spring arrives, they **pupate** and finally become butterflies. Other species in this family go through the process more quickly. They hibernate as mature caterpillars, having eaten all summer and moulted several times. In the spring, they pupate and change into butterflies. Some species in this family even hibernate as adults. In all cases, once the adult stage is reached, the cycle begins again. In many species of this family, this begins with the males patrolling their territory for receptive females.

Alpine

Fritillary

Tortoise-shell

Arctic

Food and Feeding

Butterflies from the Fritillary subfamily feed on flower nectar, as do ones from the Arctic and Alpine subfamilies. Tortoiseshell butterflies prefer tree sap or rotting fruit, but they will also consume flower nectar.

The caterpillars of the Fritillary butterflies feed on a variety of shrubs in the Arctic. They eat the leaves of blueberry, willow, dwarf birch, cranberry, and bog rosemary, and perhaps other heath plants. The Tortoiseshell caterpillars feed mostly on willow and dwarf birch in Nunavut. Caterpillars of the Arctic and Alpine subfamilies feed on grasses and sedges.

Habitat

In Nunavut, all of these butterflies are found on the tundra. Some prefer grassier areas, while others prefer dry or wet tundra.

Range

Photo by Mary Hopson

Two types of Fritillary butterflies are found only in the Low Arctic, while five members of this subfamily are found both in the High and Low Arctic (seven in total). Three Tortoiseshell butterflies are found on the Low Arctic mainland near Baker Lake, Arviat, and Rankin Inlet. Two butterflies from the Arctic subfamily are found in the High and Low Arctic, while a third species is found only in the Low Arctic. The only butterfly from the Alpine subfamily living in Nunavut is found in the Low Arctic. Together, a total of thirteen butterflies from the Nymphalidae family have been recorded in Nunavut.

Did You Know?

Caterpillars from the Arctic subfamily can survive the extreme colds of the Arctic weather by dehydrating all of their tissues and freezing solid. When it becomes warm enough in the spring, they thaw and continue their life cycles.

Snout Moths

Pyralidae

Order: Lepidoptera – Butterflies and Moths; Family: Pyralidae – Snout Moths

Members of the Pyralidae family are commonly known as snout moths in English. No specific Inuktitut names were found for this family during the research for this book. Please see the Lepidoptera account for general names for butterflies and moths. Also, if you know of names for any species in this family, please contact the Nunavut Teaching and Learning Centre.

Description

Adults:
Snout moths are very small. Their wingspans are less than eight to ten millimetres and thus they are sometimes referred to as micromoths. Their **palps** are often large and project forward, resembling snouts. They have scaled **proboscises**, long legs attached to their **thoraces**, and **tympana** (vibrating membranes like eardrums) on either side of their second abdominal segments. These organs enable moths to detect the ultrasound of insect-eating bats. Pyralid moths have elongated and triangular front wings and broad, rounded hind wings. When at rest, the wings form a triangular shape on the moth's back. Also when at rest, the antennae are pointed back over the head, while the two front legs are out in front of the moth.

Larvae:
The larvae (also known as caterpillars) of snout moths vary in colour, size, and shape.

Life Cycle

Complete Metamorphosis:
Not a lot is known about these micromoths in the Arctic. As with all butterflies and moths, the eggs are laid, and then the caterpillars emerge, **moult** several times, and **pupate**. Out of the **pupae** come the adult moths. Specifics on these small moths are yet to be recorded.

Food and Feeding

Some species are known to feed on grasses as larvae. Other specific information about their feeding habits is lacking.

Habitat

These moths live on both dry and wet tundra.

Range

There are less than one dozen of these small moths across the Arctic and probably even fewer in Nunavut. One species does, however, reach the High Arctic.

Did You Know?

In the South, species of the Pyralidae family are serious pests, living on and eating certain crops. For example, the larvae of one species bore into corn to feed.

Photo by Johannes F. Skaftason

Photo by Johannes F. Skaftason

Order: *Lepidoptera*

Looper Moths

Geometridae

Order: Lepidoptera – Butterflies and Moths; Family: Geometridae – Looper Moths

Members of the Geometridae family are commonly known as looper moths, geometer moths, or inchworms in English. No specific Inuktitut names were found for this family during the research for this book. Please see the Lepidoptera account for general names for butterflies and moths. Also, if you know of names for any species in this family, please contact the Nunavut Teaching and Learning Centre.

Description

Adults:
Looper moths are small to medium in size, and have slender bodies. The male moths in this family often have feathered antennae. Members of both sexes have **tympana** on the first segments of either side of their **abdomens**. They are bare, shiny, drum-like plates that help looper moths "hear." Looper moths have broad wings with similar patterns on both the hind- and front wings. Fine, wavy lines on the wings help to create a camouflage pattern that allows the moths to blend into their surroundings.

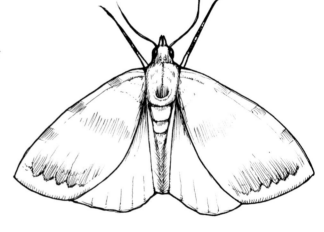

Larvae:
The larvae, also known as caterpillars, are slender and usually hairless. They have distinct heads and can be green, grey, or brown, which helps them to blend in with leaves or branches. Looper moth caterpillars are commonly called inchworms because of their style of locomotion. Each caterpillar has three pairs of true legs at the front end, no legs in the middle of the body, and two pairs of **prolegs** at the hind end. To move forward, it stretches out its body, then moves the hind end up to the front end, forming an upside-down U shape. Then it stretches out the front end again. The caterpillar looks like it is measuring small segments as it goes, which is how it got the name inchworm.

Photo by Johannes F. Skaftason

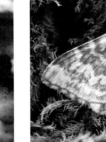

Photo by Johannes F. Skaftason

Life Cycle

Complete Metamorphosis:
The life cycles of looper moths vary, since some **overwinter** as pupae, while others overwinter as eggs. In both cases, the moths mate, lay eggs, go through several **instars** as caterpillars, and then **pupate**, before finally emerging as adult moths.

Food and Feeding

In the South, most caterpillars from this family feed on the leaves of trees or shrubs. Most are also host-specific, meaning that they eat one type of plant only. In the High Arctic, it is known that at least one species of caterpillar feeds on a variety of low plants. Almost all adult looper moths lack mouthparts and therefore do not feed.

Habitat

These moths can be found on dry tundra around potential food sources for their larvae.

Range

There are sixteen species of looper moths in Nunavut, two of which have been recorded in the High Arctic.

Did You Know?

Some caterpillars, to confuse **predators**, drop to the ground when threatened and curve their bodies into circular shapes. Since the caterpillars look completely different as round objects, the predators may not recognize them as food any longer.

Order: *Lepidoptera*

Tussock Moths

Lymantridae

Order: Lepidoptera – Butterflies and Moths; Family: Lymantridae – Tussock Moths

Members of the Lymantridae family are commonly known as tussock moths in English. No specific Inuktitut names were found for members of this family during the research for this book. Please see the Lepidoptera account for general names for butterflies and moths. Also, if you know of a specific name for any species in this family, please contact the Nunavut Teaching and Learning Centre.

Description

Adults:
These medium-sized moths have wingspans of twenty to seventy millimetres. They are stout-bodied and hairy, and are brown to white in colour. Males have feathery antennae and both sexes lack **ocelli**. They have reduced or absent **proboscises**. Males have a **tympanum** on either side of their bodies. In some species, the females are wingless or have reduced wings, while males of all species have broad wings.

Tussock moths hold their wings like a roof over their bodies when at rest.

Larvae:
Tussock moths have very distinctive larvae caterpillars called wooly-bear caterpillars. They are hairy with a pair of pencil-like hair tufts at the head and one single tuft at the tail end. Some species also have four short, thick tufts on their backs. They are slender but appear thick because of all of the tufts. They are sometimes brightly coloured.

Life Cycle

Complete Metamorphosis:
Some females recover hair from their **cocoons** to camouflage and protect their eggs. Other females cover their eggs with a froth that hardens, while still others collect material and stick it to their eggs as camouflage. From the eggs emerge the caterpillars. The caterpillars eat and grow, **moulting** several times before they **pupate**. When they are ready to pupate, the caterpillars weave cocoons, which include some of the hair from their bodies, and hide under the cover of leaves.

The summer season can be very short and cold in the High Arctic, and it was once thought that these moths took between fourteen and twenty years to complete one life cycle. More recent science has indicated that tussock moths likely take about seven years to complete this cycle. Members of one species found in the High Arctic, *Gynaephora groenlandica*, stop feeding at the end of June and find hiding places until the following spring. Part of the reason for this may be to reduce the chance of being captured by the predaceous parasitoids, such as wasps and flies, that appear at about this time.

Photo by Doug Macaulay

Another reason could be the subtle cooling caused by the change of the angle of the sun at the end of June—these caterpillars are more highly dependent on sunlight to stay warm than other species. These caterpillars spend 60% of their time basking in the sun, while other caterpillars spend most of their time eating. A third reason could be that the chemical composition of Arctic willow changes at this time, producing less of the essential nutrients needed by the wooly-bear caterpillars.

Food and Feeding

Most of the adults in this family do not feed. The caterpillars eat leaves and their favourite plant in Nunavut is the Arctic willow.

Habitat

Tussock moths and caterpillars can be found on the tundra near their favourite foods.

Range

Two species from this family occur in Nunavut. They are found in both the High and Low Arctic.

Did You Know?

The hairs of the caterpillars in this family break off very easily and cause skin irritations in many people. That is why the adults may protect their eggs with these hairs and also incorporate them into their cocoons.

Order: *Lepidoptera*

Owlet Moths

Noctuidae

Order: Lepidoptera – Butterflies and Moths; Family: Noctuidae – Owlet Moths

Species of the Noctuidae family are commonly known as owlet moths in English. No specific Inuktitut names were found for this family during the research for this book. Please see the Lepidoptera account for general names for butterflies and moths. Also, if you know of names for any species in this family, please contact the Nunavut Teaching and Learning Centre.

Noctuidae is the largest family in the Lepidoptera order.

Description

Adults:
Owlet moths have wingspans of twenty to forty millimetres. They almost always have **ocelli** in addition to their **compound eyes**. Their compound eyes pick up minute light and shine brightly, hence the name "owlet moths." They have well-developed **proboscises** and slender, threadlike antennae, which are never feathery.

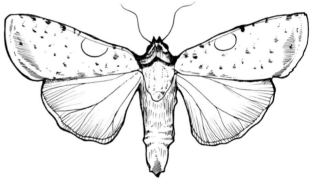

Their **palps** extend to the middle of their faces and sometimes beyond. On each side of their **thoraces**, owlet moths have a prominent **tympanum**. Noctuid moth wings are mottled, dull grey or brown. The front wings are narrow, while the hindwings are broad. When at rest, the wings are held over the body, flat and roof-like.

Larvae:
The dull-coloured larva, commonly known as a caterpillar, is usually smooth and without obvious hairs. Most have five pairs of **prolegs**, although some only have three pairs. The owlet moth caterpillar moves similarly to the geometer looper moth caterpillar (Geometridae family)—by inching its way along.

Life Cycle

Complete Metamorphosis:
Like other members of the Lepidoptera order, owlet moths lay eggs, which then hatch into caterpillars. The caterpillars go through several **moults** as they eat and grow. Most owlet moths **pupate** in the soil, but some also pupate in plant cavities or in **cocoons** under leaves. They tend to **overwinter** as larvae, pupate in the summer, and then finally become adult moths and begin the cycle again.

Food and Feeding

The caterpillars feed on plants, dead leaves, lichen, and fungi. Some eat their way through leaves, making noticeable trails. Others bore into stems and leaves. Some owlet moth caterpillars live in the top layer of soil and cut off new plant shoots at the base to eat. These last caterpillars are called cutworms and are very familiar to gardeners in the South. Many species are not host-specific, choosing to eat a variety of plants. The adults feed mostly on nectar from flowers, but some feed on ripe fruit.

Photo by Johannes F. Skaftason

Habitat

As adults, owlet moths can be found flying on the tundra near flowers, and as caterpillars, they can be found hiding among low plants.

Range

Twenty-four species of Noctuid moths have been identified in Nunavut, at least five of which can be found in both the High and Low Arctic.

Did You Know?

Using their tympana, owlet moths can detect the sonar systems that bats employ when hunting prey. With these "ears," owlet moths can tell how far away the bats are and which direction they are coming from. That way, they can avoid being eaten. This ability is not particularly useful in Nunavut, since bats only make it to the southernmost parts of the territory.

Caddis Flies

Order: Trichoptera – Caddis flies

Members of the Trichoptera order are commonly known as caddis flies in English. No Inuktitut names were found for this order during the research for this book. If you know of any, please contact the Nunavut Teaching and Learning Centre.

Caddis flies are closely related to butterflies and moths. The name Trichoptera comes from the Greek words *trich*, meaning hairy, and *ptera*, meaning wings.

Description

Adults:
Caddis flies are long, slender, and moth-like. They vary from 1.5 to twenty-five millimetres in length. Their threadlike antennae are usually as long as, or longer than, their bodies. Although their **palps** are well-developed, their **mandibles** are greatly reduced. Caddis flies have highly functional **compound eyes**. Although they have four **membranous** wings, they are poor fliers.

Photo by Carolyn Mallory

Their hindwings are a little shorter than their front wings, and all their wings are quite hairy. When caddis flies are at rest, they hold their wings like tents over their bodies. They have long, slender legs with two strong claws. Caddis flies are generally dull-coloured.

Larvae:
Caddis fly larvae are sometimes referred to as caddis worms. They are caterpillar-like, with well-**sclerotized** heads and **thoraces**. They have **ocelli** and chewing mouthparts. They also have small antennae. Caddis worms have three pairs of walking legs on the thorax and one pair of **prolegs** at the tip of the **abdomen**. Most larvae have external **gills**.

Life Cycle

Complete Metamorphosis:
Female caddis flies lay small, round, pale eggs that are blue, green, or white. The eggs are usually in a jellied mass or in strings. The female attaches them to rocks, plants, or debris in or near water. The eggs hatch into larvae which, in many species, make cases out of silk and grains of sand, shells, or small pieces of plants. They live inside these cases, which protect and camouflage them from **predators**. In other words, the larvae make small

houses that they drag around with them! Some larvae do not make cases, but instead make silk nets that they stretch between underwater rocks or plants. A few larvae do neither.

The larvae go through five **instars** before they **pupate**. Each time they shed their skin and grow, they must reconstruct their cases. Larvae with cases will pupate inside of them, while larvae without cases will make silk **cocoons**. When ready to pupate, larvae attach their cases or cocoons to objects in the water. The pupae have biting mandibles to chew through the case or cocoon when the rest period is over. They then swim to the surface and climb onto land, where they emerge with wings. Many species exhibit synchronous emergence, which means that they all emerge from their pupae at about the same time. This makes finding a mate very easy. Adults live for about two weeks before the cycle begins again.

Photo by James Lindsey

Food and Feeding

Most case- or net-building larvae feed on plants, but some are **predaceous**. The larvae that construct nets catch their food by trapping it in those nets. Adults rarely eat.

Habitat

The larvae live under water in quickly flowing streams, still lakes, or stagnant ponds. It is thought that they prefer cool, well-aerated water. Adults live near the same bodies of water, as the females will deposit their eggs there.

Range

Most of the caddis flies in Nunavut are found in the Low Arctic and belong to the Limnephilidae family. One member of this family has also been found in the High Arctic. There are approximately fifteen species known to live in the Arctic.

Did You Know?

Adult female caddis flies can stay underwater for more than thirty minutes. This is possible because the hairs on their bodies and wings create air bubbles around them, which more or less act like gills. They breathe in oxygen and diffuse carbon dioxide into the water through the thin film of the air bubbles.

Bees, Wasps, Ants, and Sawflies

Order: Hymenoptera – Bees, Wasps, Ants, and Sawflies

Members of the Hymenoptera order are commonly known as bees, wasps, ants, and sawflies in English. For Inuktitut names of these insects, please see each specific account.

There are two suborders in the Hymenoptera order. The Apocrita suborder contains bees, wasps, and ants, and the Symphyta suborder contains sawflies.

Description

Adults:
Hymenopterans are typically between two and thirty millimetres long. They have hard bodies and range from nearly hairless to very hairy. The antennae of hymenopterans have at least ten segments. Their mouthparts are mostly for chewing, but members of some families have modified **labia** (lower lips) and **maxillae** (paired mouthpart structure beside the jaw) that form tongue-like, sucking structures. Hymenopterans have two pairs of membranous wings. The front wings are a little larger than the hindwings. Each hymenopteran possesses a row of tiny hooks that attach the front wings to the hindwings when in flight. These are called **hamuli**. The wings of insects in this order have relatively few veins, with the exception of the sawflies', which have complex venation. Some adults in this order are wingless. Bees, wasps, and ants have constricted waists, while sawflies do not. The females have well-developed **ovipositors** (egg-laying apparati) which can sometimes be modified into stingers. Only females can sting. In the sawflies, the ovipositor is saw-like, and serves to insert eggs into plant tissue.

Larvae:
Most larvae belonging to the Apocrita suborder (bees, wasps, and ants) are thin and legless, with well-developed heads and mouthparts adapted for chewing. They have no eyes. The exception is the parasitc wasps (Braconidae), which do not have heads, eyes, or legs. The larvae in the Symphata suborder (sawflies) are more like caterpillars. They have six fully developed legs and six or more **prolegs**. They also have one pair of **ocelli**.

Life Cycle

Complete Metamorphosis:
The life cycles of insects in this order are somewhat variable. Sawflies lay their eggs on or in plant tissue. They use their saw-like ovipositors to cut holes in plant tissue and insert their eggs. The sawfly larvae hatch into caterpillar-like young and they eat and grow, **moulting** several times before resting and transforming into adults during the **pupal** stage.

Parasitoid wasps (Braconidae) lay their eggs on or near their **hosts**. The eggs hatch inside or near the hosts. These hosts can be anything from larvae to sawflies, or even the eggs of another insect. Some wasps even parasitize other parasitoids—they live inside parasitoids that live inside other hosts! The hosts are eventually killed by the developing young. The wasp larvae feed on the eggs, larvae, and bodies of the adult hosts, and when the wasp larvae are ready to **pupate**, the host dies. A new parasitic wasp adult emerges and is ready to begin the cycle again.

Photo by Johannes F. Skaftason

With true wasps (those that are not parasitoids), bees, and ants, the eggs are laid by a **queen** inside a nest. Most of these insects are social, which means they all work together to benefit the group. The larvae do not have eyes or legs. Because the workers bring them food, these larvae have no need to move or see. Inside the nest, the larvae grow and **moult** until it is time to pupate. New adults emerge from the pupae and the cycle begins again.

One of the very interesting facts about the Hymenoptera order is that the females develop from fertilized eggs, while the males develop from unfertilized eggs.

Food and Feeding

Parasitic wasp larvae feed on other insects or insect eggs. Other larvae are **herbivores**, eating a variety of plants. Still others feed on pollen. Adults in this order also eat a variety of foods. Some feed on flower nectar or pollen, while some eat **honeydew** or fungus. Still other adults are **predators**, hunting and feeding on other insects, while some adults do not feed at all.

Habitat

Many hymenopterans can be found around plants on the tundra. Bees also need underground nesting areas, often under rocks. In Nunavut, some of the nest-building wasps need to be close to human habitations in order to obtain their nest-making materials.

Range

Over 160 named hymenopteran species have been found in Arctic North America. At least eighty hymenopteran species occur in the High Arctic, though not all of them have been named.

Did You Know?

Despite common perception, the order Hymenoptera is quite well-represented in the Arctic. Most people see the occasional bee and nothing else. A number of sawflies and many parasitoid wasps inhabit the tundra, but they tend to go unnoticed.

Traditional Knowledge

From Taloyoak: Ellen Ittunga interviewing Mary Ittunga, July 2006

When I was a child, we were told never to play with or torture insects or any living being. If we're not going to eat them, we have to leave them alone. Today, children are not taught this traditional value anymore. You see children running after rodents or birds and throwing stones at them. We were always taught not to kill any animals or birds if we had no intention of eating them. Every living being suffers just like any other. They say it can always come back to haunt us. I believe this, because when one of my brothers was a child, he was pouring water into a weasel's habitat. There was no doubt that the weasel was suffering and going to drown. Later that night, my brother [had a nightmare in which] his children [were] chased by a giant weasel. We must teach these kind of traditional morals to our children.

Order: *Hymenoptera*

Common Sawflies

Tenthredinidae

Suborder: Symphyta; Family: Tenthredinidae – Common Sawflies

Members of the Tenthredinidae family are also known as common sawflies in English. No Inuktitut names were found for members of this family during the research for this book. If you know of any, please contact the Nunavut Teaching and Learning Centre.

Tenthredinidae is the largest family of sawflies.

Description

Adults:
Common sawflies are generally between three and twenty millimetres long. The adults are wasp-like but, unlike wasps, they do not have cinched waists. They are hard-bodied and shiny. Common sawflies are mostly black or brown, but some can be brightly patterned. They have threadlike antennae with at least nine segments. The antennae are often **clubbed** or comb-like. The female common sawflies have saw-like **ovipositors**.

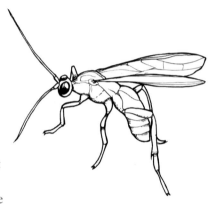

Larvae:
Common sawfly larvae look like caterpillars. They have well-developed heads with chewing mouthparts. The larvae have three pairs of legs as well as at least five pairs of **prolegs** on their **abdomens**. The larvae are hairless. They have one pair of **ocelli**.

Life Cycle

Complete Metamorphosis:
The sawfly female has an ovipositor resembling a saw blade, with which she can make incisions into leaves or stems and then deposit the eggs. Once the eggs hatch, the small larvae emerge and start feeding, often in groups. They go through several **instars** before spinning **cocoons** in which they will later **pupate**. Many sawflies **overwinter** as pupae and become adults in the spring, thus starting the cycle again.

Food and Feeding

Most larvae of the Tenthredinidae family eat leaves. There are a few species, however, that form **galls** (abnormal growths of plant tissues) on leaves and stems and feed inside these galls. In the Arctic, most sawflies eat willow leaves. Some species of sawflies have also been found eating blueberry, cranberry, or birch leaves, and even some sedges. The adults do not eat.

Habitat

You will undoubtedly find sawfly larvae on the tundra, feeding on their favourite plants. Adults are generally found on foliage or flowers.

Range

Thirty-three species have been found in Arctic North America found so far. At least three of these species occur in the High Arctic of Nunavut.

Did You Know?

Sawfly larvae often feed on leaves in groups. With three legs on one side of the leaf margin and three legs on the other, the larvae squish close to one another to eat. The rest of their bodies hang off the leaves and form S-shapes. Keep your eyes peeled!

Photos by Olivia Mallory

Order: *Hymenoptera*

Braconid Wasps

Braconidae

Suborder: Apocrita, division Parasitica; Family: Braconidae – Braconid Wasps

These insects are commonly known as braconid wasps in English. Elders in Clyde River have identified small wasps with long ovipositors, similar to those of braconid wasps, as *kittuqsaq* in Inuktitut.

Braconid wasps are mostly **parasitoids**. As larvae, they live inside their **hosts**, consuming and eventually killing them.

Description

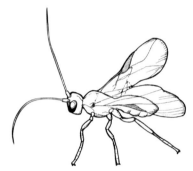

Adults:
Braconid wasps are between two and fifteen millimetres long. They are usually black or brown in colour, although some species mimic bees in their colouration. Braconid wasps have slender, unmarked antennae that sometimes have more than sixteen segments. Braconid wasps have elongated, thin bodies and narrow waists. The females often have very long **ovipositors**.

The wings of braconid wasps have one or no **recurrent veins**. A recurrent vein is one that turns back, reversing its direction. This is important to note because members of the Ichneumonidae family of wasps have two recurrent veins and this distinction allows scientists to distinguish between species in the Ichneumonidae and Braconidae families. A further distinction is the two window-like, closed **cells** (spaces between the veins) in the braconid wasp wing. These two cells are about the same size and are located under the dark, thickened edge of the wing, also called the wing margin.

Larvae:
Braconid larvae are cream-coloured, legless, **maggot**-like creatures with chewing mouthparts.

Life Cycle

Complete Metamorphosis:
The female braconid wasp lays her eggs through her long ovipositor into the host or the host's eggs. The larvae hatch inside the host and begin to consume it from the inside, or, in some cases, from the outside. The larvae **moult** five times before

pupating. Some pupae remain inside their hosts until they are mature. Other species spin **cocoons** which they attach to the outsides of their hosts. Still others spin cocoons and pupate away from their hosts. The adults emerge from their hosts and the cycle begins again.

Food and Feeding

Adults feed on nectar, **honeydew**, and pollen. Most braconids are parasitoids as larvae. They most often parasitize fly, butterfly, and beetle larvae, although they have been known to parasitize adults and plants by forming galls. In the Arctic, the hosts that braconids parasitize are not well-known, but it is thought that most are butterflies.

Habitat

Braconid larvae live inside their hosts. The adults live on the tundra near flowers on which they can feed.

Range

A small number of braconid wasps have been reported in the Arctic. However, this is a large family, and it is expected that many more braconids that live there remain to be discovered and identified. Several unnamed species from this family have been recorded in the High Arctic.

Did You Know?

Different species of female braconid wasps have ovipositors of different lengths. The size of the ovipositor is directly related to the host species into which the wasps' deposit their eggs. For example, some braconid wasps have long ovipositors because the caterpillars they parasitize have long spines or hairs. The long ovipositors allow the female wasps to land on the caterpillar and reach past the hairs and spines, inserting her eggs into its body.

Photo by James Lindsey

Order: *Hymenoptera*

125

Ichneumon Wasps

Ichneumonidae

Suborder: Apocrita, division Parasitica; Family: Ichneumonidae – Parasitic Wasps

Members of the Ichneumonidae family are commonly called parasitic or ichneumonid wasps in English. Elders in Clyde River identified small wasps with long ovipositors like ichneumonid (and braconid) wasps as *kittuqsaq*.

Ichneumonidae is the largest family in the Hymenoptera order.

Description

Adults:
Ichneumonid wasps vary greatly in shape, size, and colouration. Some are uniformly coloured (anywhere from yellow to black) while others are brightly patterned (black and brown or black and yellow). They range from three to seventy-three millimetres in length. Ichneumonid wasps are larger and slenderer than braconid wasps. Another distinction between these two families lies in the fact that the ichneumonid wasp's **abdomen** is longer than its head and **thorax** combined. This is not true for the braconid wasp. Ichneumonid wasps have long antennae with sixteen or more segments. These antennae can be yellowish or white in the middle segments. As with the other members of the Hymenoptera order, ichneumonid wasps have narrow waists. They have long, threadlike **ovipositors** that are permanently sticking out of the ends of their abdomens. On their wings, ichneumonid wasps have two **recurrent veins**.

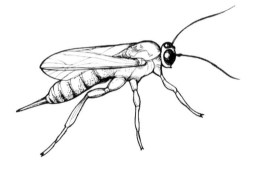

Larvae:
The larvae are **grub**-like. They have not been described very thoroughly in available literature.

Life Cycle

Complete Metamorphosis:
Most ichneumonid wasps are **parasitoids**, living and feeding on caterpillars and sawflies in both their larval and **pupal** states. Some also attack spider egg sacs. Once mating has occurred, the female must locate a **host** in which to lay her eggs. Many ichneumonid wasps do this first by locating their host's food source, then searching out the host. These wasps can locate host larvae under leaves or bark by tapping the plant surface with their antennae.

When the host is found, the female ichneumonid will inject her eggs into the host using her ovipositor. Sometimes, if the host is confined to a small space, such as inside a gall or some other part of a plant, the eggs are injected inside the space so they can feed on the host from the outside. If the host is injected directly, the larvae feed on the host from the inside. The larvae go through several **moults** before **pupating**. Finally, they emerge from their dead host to begin the cycle again.

Photo by Johannes F. Skaftason

Food and Feeding

In the Arctic, ichneumonid wasp larvae mainly parasitize butterfly, fly, and sawfly larvae. The adults drink nectar and water.

Habitat

Ichneumonid wasps can be found on the tundra near their host species.

Range

Over 160 species in the family Ichneumonidae have been reported from Arctic North America. Many species that have been collected remain unidentified and unnamed as well. At least fifty species reach the High Arctic, and many of these are as yet unnamed.

Did You Know?

Ichneumonid wasps do not sting in defense, though you might be frightened by their ovipositors, which can look like long stingers. You can distinguish ichneumonid wasps from stinging wasps because their antennae have sixteen or more segments, while the antennae of stinging wasps have thirteen segments or less.

Photo by Johannes F. Skaftason

Social Wasps

Vespidae

Suborder: Apocrita; Family: Vespidae – Vespid Wasps

Members of this family are commonly known as vespid wasps, yellowjackets, or hornets in English. In Clyde River, vespid wasps are called *iguttaq* and *aanngiq*. One elder in Iqaluit referred to a photo of a vespid wasp as an *iguptaujaq* in Inuktitut, while in Baker Lake, the vespid wasp is known as an *aabjiq*.

Description

Adults:
Vespid wasps are medium to large wasps ranging from ten to thirty centimetres long. They range in colour from brown to black and are banded with yellow or white. Most vespid wasps have notched eyes, meaning that their eyes are not completely round. The eyes have indentations on their inner sides where the antennae arise. The first segment of the antennae can be yellow. Vespid wasps have very long first discoidal cells in their front wings, which is another distinguishing characteristic of theirs. This cell is almost half the length of the wing. When the vespid wasp is at rest, its wings are folded lengthwise over its **abdomen**, appearing to be pleated. As with all wasps, its waist is narrow.

Larvae:
The larvae are **grub**-like, and do not have legs or eyes—they spend their entire lives inside their nests and therefore do not need them.

Life Cycle

Complete Metamorphosis:
A fertilized **queen** that has **overwintered** begins a new colony in the spring. She forages for food and builds a small nest in which to lay her eggs. Each egg is laid in a six-sided cell. The queen feeds the larvae once they hatch. The larvae shed their skins several times before they **pupate**. The adults that emerge from this first laying are workers, all of which are sterile females, incapable of laying eggs. These workers build larger nests and feed the next group of larvae as the queen continues to lay eggs. When the end of the season draws near, the queen produces both fertile females and males. The new females

Photo by Carolyn Mallory

mate and then overwinter under leaves. All the other members of the colony die. The new females will begin the cycle again in the spring by each building their own colony. Note: One divergence from this pattern is seen in *Vespula austriaca*, a vespid wasp that is a social **parasite**. The queen enters a nest, kills the queen inside, and uses the workers from that nest to raise her own larvae.

Food and Feeding

The adults catch and chew up flies, caterpillars, and other **arthropods** to feed their larvae. In the process of doing this, the adults gain some nourishment as well. The adults also drink the saliva of their larvae.

Habitat

Vespid wasps need to live where they can collect old wood or cardboard to make their nests, which in Nunavut means living near human habitation.

Photo by Johannes F. Skaftason

Range

Vespid wasp nests have been found around Iqaluit and elders have reported seeing yellowjackets in both Baker Lake and Iqaluit. Two species from the subfamily Vespinae have been reported in the Low Arctic of Nunavut: the Arctic yellowjacket (*Dolichovespula norwegica*) and the Boreal yellowjacket (*Vespula austriaca*). *V. austriaca* has also been found in Nunavut. It is a social parasite of *Vespula rufa* and, although *V. rufa* has not yet been captured in Nunavut, it is expected to be there because its parasite is.

Did You Know?

Male yellowjacket wasps hatch from unfertilized eggs. They do not have stingers. Female yellowjacket wasps have stingers, and defend their nests with them.

Traditional Knowledge

From Baker Lake: Brenda Qiyuk interviewing Silas Aittuq, July 2006

BQ: *[Looking at photo of yellowjacket]* Are there any stories, superstitions, or uses associated with this bug?

SA: *I have heard that they would catch this bug, put something around it, and leave it on the ground. I do not know if this [was] a joke but they said that they did that so that there would be caribou close by.*

Bees

Apidae

Suborder: Apocrita
Family: Apidae – Cuckoo, Carpenter, Digger, Bumble-, and Honey Bees
Subfamily: Apinae – Digger, Bumble-, and Honey Bees

In Baker Lake and Taloyoak, the Inuktitut name for a bee is *igutsaq* (*igupsaq*). In Iqaluit, Clyde River, and Igloolik, the name is very similar: *iguttaq* or *iguptaq*.

The most common type of bee found in Nunavut is the bumblebee, which belongs to the subfamily Apinae. The information that follows is about bumblebees in this subfamily.

Description

Adults:
Bumble bees can be fifteen to twenty-five millimetres long. The **queens** are larger than the workers. They are robust and have branched or feathery hair. The thick hair is called "the pile" and it can range in colour from yellow to orange, with black stripes. Bumblebees have two large **compound eyes** and three **ocelli**. They have two long, straight antennae with twelve or thirteen segments (males have one more segment than females). Bumblebees have long tongues that enable them to reach the nectar in flowers. Their tongues are hairy on the end, better allowing the bumblebees to suck up the nectar. When not in use, their tongues are folded up under their heads and protected by sheaths made with the **palps** and the **maxillae**. Bumblebees have four wings, just like all other bees. The front wings are larger than the hindwings. The hind legs are equipped with comb- or brush-like hairs on the inside that help transfer the pollen stuck there to the pollen basket, which is located on the outside of the hind legs of female bumblebees. Female bumblebees have stingers at the ends of their **abdomens**, while the males do not.

Larvae:
Bumblebee larvae look like **grubs**. They are soft-bodied, white, and legless. The larval head is hardened and brown.

Life Cycle

Complete Metamorphosis:
The queens that have mated by the end of the summer then **overwinter** in separate, sheltered locations. In the spring, the queens each start a colony. Each queen collects pollen and nectar and then finds a suitable nesting place, be it an old rodent burrow or a tussock of grass. Next, she constructs a wax honey pot that she fills with honey. She also collects pollen, deposits it on the floor of the nest, and lays some eggs on it.

Photo by Carolyn Mallory

When the larvae hatch, the queen tends to them, feeding them honey and pollen. They go through four **instars**. At the end of the fourth instar, the larvae spin silk **cocoons** in which they **pupate**. The adult bees emerge and take over the duties of the queen, such as foraging for food and keeping the nest clean. All of the adults produced are infertile females. The queen continues to lay eggs, and the colony grows over the summer. In the Arctic, only one more brood is produced, due to the short summer. Southern colonies become much larger. In the second brood, both fertile females and drones (males) are produced so that the cycle can once again begin in the spring with the overwintering, mated queens. The old queens die shortly after producing the second brood, and the other bees die as well. Only the mated queens overwinter.

Of the two most common Nunavut species of bumblebees, *Bombus polaris* follows the life cycle as described above, while *Bombus hyperboreus* follows the life cycle of a **parasite**. Once a *B. hyperboreus* queen emerges in the spring, she looks for an active *B. polaris* nest. She attacks and kills the queen, then enslaves the female workers. This new queen then produces more queens and drones, which the unrelated worker bees care for. *Bombus hyperboreus* is interesting because it is not a true parasite, since farther south it behaves exactly like other bees, making its own nest and caring for its young.

Order: *Hymenoptera*

Food and Feeding

Adult bumblebees eat pollen and drink nectar, both from flowers. Larvae are fed pollen, nectar, and honey.

Habitat

Adult bumblebees need flowers to survive, so they are often found flying around or resting on flowers. Larvae never leave their nests, which may be abandoned lemming holes, other types of holes, or tussocks of grass.

Range

Less than half a dozen bumblebee species have been positively identified in Nunavut; however, the collecting of bees in the territory has not been very extensive, and more are thought to live there. Three species live in the High Arctic.

Did You Know?

The queen bee needs to mate only once to have enough sperm to last her whole life. The container in which she stores the sperm after mating is called a **spermatheca**. As an egg goes down the **oviduct**, the queen decides if the egg will be fertilized (to produce a female) or not (to produce a male) by allowing sperm to exit the spermatheca at the correct moment.

Photo by Carolyn Mallory

Photo by Carolyn Mallory

Traditional Knowledge

From Iqaluit: Sylvia Inuaraq interviewing Mary Ell, July 2006

The bees are different sizes. Both children and adults are scared of them. It is said that if a bee stings you and the stinger gets stuck in your skin, you might get sick.

From Clyde River: Rebecca Hainu interviewing Kalluk Palituq, July 2006

Kalluk: *No, people should not use insects to scare other people—especially [not]* iguttaq. *It is known that* iguttaq *will seek revenge on the teaser.*

Rebecca: *Really. Why do people chant "qaariaq, qaariaq" towards* iguttaq?

Kalluk: *We are not supposed to use them to scare people. We chant* qaariaq *in our attempt to scare the* iguttaq *away. [Qaariaq is a threat to pop the iguttaq].*

Order: *Hymenoptera*

Class Arachnida

Arachnids

■ Araneae – True Spiders 138
    Lycosidae – Wolf Spiders 140
    Philodromidae – Crab Spiders 142
    Lyniphiidae – Sheet Web
    and Dwarf Spiders 144

Subclass Acari – Mites 146
    Some Interesting Mites 148

Arachnida

Arachnids

Class: Arachnida

Members of this class are commonly known as arachnids in English. No Inuktitut names for this class were found during the research for this book. If you know of any, please contact the Nunavut Teaching and Learning Centre.

This class has ten orders, including spiders, ticks and mites, scorpions, and others. However, only species from the order Araneae (true spiders) and the subclass Acari (ticks and mites) are found in Nunavut.

Description

Adults:
There are several significant differences between arachnids and insects. Arachnids have no wings or antennae and they have eight legs, unlike insects, which have six. Arachnids only have two distinct body parts, whereas insects have three. The two distinct body parts of most arachnids are the **cephalothorax** (fused head and thorax) and the **abdomen**. There are some exceptions among the arachnids, as mites and ticks have fused body parts and therefore seem to have only one. Arachnids have two pairs of appendages around the mouth: **chelicerae** (jaw-like appendages) and **pedipalps** (feeler-like parts). Most arachnids have **carapaces** (protective coverings) on the hind parts of their cephalothoraces. Also, most arachnids have four pairs of eyes. They generally have no abdominal appendages, except for spiders, which have silk-spinning **spinnerets**, and scorpions, which have comb-like sensors. Arachnids breathe using **book lungs** (sac-like respiratory organs that have several folds arranged like the pages of a book) or **tracheae** (tubes that transmit oxygen to all parts of the body) or both.

Nymphs:
Arachnid nymphs are just smaller versions of the adults.

Life Cycle

Incomplete Metamorphosis:
Arachnids go through incomplete metamorphosis, which means they lay eggs, out of which hatch nymphs that resemble the adults, only smaller. Later they become sexually mature adults. Arachnids do not go through **pupal stages** during which complete changes take place. Scorpions are an exception because they give birth to live young. For specific information about the life cycles of arachnids, please see the order and family accounts.

Food and Feeding

Arachnids are **predators**, capturing and eating their prey, except for a very small number of plant-eating mites. Most arachnids can only consume liquids, but a few species of mites can ingest solid foods. Arachnids pierce the bodies of their prey (or, in the case of plant-eating mites, the skins of the plants) and then let flow their salivary juices, which contain enzymes that help begin the digestive process. The prey's tissues break down and the arachnids suck up the "soup" that is formed.

Habitat

Arachnids are found in every possible habitat. Although they are mostly **terrestrial**, there are a few species that live in both marine and fresh waters.

Range

Of the ten orders in the class Arachnida, only one order is present in Nunavut: Araneae, or true spiders. However, many species of mites form the subclass Acari are present in Nunavut. The other orders, not found in Nunavut, include organisms such as daddy longlegs spiders, scorpions, and false scorpions. The arachnids of Nunavut have been poorly studied thus far, so it is difficult to estimate how many species actually live there.

Did You Know?

Thousands of species of arachnids have been identified worldwide. In fact, there are over 45,000 species of mites and ticks alone.

True Spiders

Order: Araneae – True Spiders

Members of the Araneae order are commonly known as spiders in English. In Baker Lake, a true spider is an *aasivak* or a *nanuujaq*. In Iqaluit, spiders are known as *aasivak*, *nanuujaq*, *nivingasuuq*, *nigjuk*, or *nigjuarjak*. In Taloyoak and Clyde River, a spider is an *aasivak*. The terms *nigjuk* and *aasivak* are also used in Igloolik.

True spiders are the largest group of arachnids.

Description

Adults:
Spiders have two body sections: the **cephalothorax** and the **abdomen**. They can range from 0.4 millimetres to nine centimetres in length. The back or top of the cephalothorax has a hard, protective covering called the **carapace**. Most spiders have eight **ocelli**. The number and arrangement of the eyes is important for identifying the different families of spiders. Spiders have poor eyesight, so they mostly use vibrations to "see" things. Spiders have two pairs of appendages around the mouth. The first pair is the **chelicerae**, which end in fangs. Venom is produced there and used to subdue prey. The second pair is the **pedipalps**, which are larger in males as they are used to hold the female during mating. Spiders do not have antennae. At the rear ends of their **abdomens**, spiders have one to four pairs of finger-like organs called **spinnerets**. These produce silk for web-making, descending or ascending from heights, and egg sac construction. Spiders have four pairs of walking legs, with seven segments each. Some spiders are brightly coloured, while others are camouflaged.

Life Cycle

Incomplete Metamorphosis:
Female spiders can produce hundreds, even thousands, of eggs. These eggs are usually held in silk cases or egg sacs that keep them dry and at a good temperature. Some females protect their egg sacs by standing guard over them, while other females carry the egg sacs at the tips of their abdomens or in their jaws. Some spiders offer no protection except camouflage for their egg sacs. Once the **spiderlings** hatch, they may be left on their own, or the mother may provide some degree of care. True spiderlings, for example, ride on their mothers' backs for the first week of their lives. Other species feed their young. Spiderlings **moult** four to twelve times, depending on the species, before becoming sexually mature adults.

Male spiders are generally much smaller than females. This is thought to help prevent the females from feeling threatened and eating the males. Males have also evolved elaborate mating rituals, again to avoid being eaten by the females. The mating dances may last only a few seconds, or may go on for a few hours. Before searching for a mate, the male spider releases sperm from his genital opening, which he then sucks up into his modified pedipalp. He is then ready to find a female, transfer the sperm to her genital opening, and begin the cycle again.

Food and Feeding

All spiders are **predators**, preying mostly on insects and other spiders. Spiders can be divided into two broad groups in terms of their feeding strategies: web builders and hunters. Web builders lie in wait and trap their prey, while hunters ambush their prey without the use of webs. Digestion in spiders begins outside of their bodies. They vomit digestive fluids onto their paralyzed or tied-up prey. They then suck up the liquid resulting from this process. Spiders have slow metabolism and can last for months without food.

Habitat

Spiders are found in all terrestrial environments except ice caps. In Nunavut, they are found in houses and on the tundra, where they may be harder to spot because they are well-camouflaged.

Range

Spiders are found all across Nunavut. The most common families are Linyphiidae, Lycosidae, and Philodromidae. The farther north you travel, the fewer species you will see.

Did You Know?

Although spiders consume only a liquid diet, they are capable of digesting their own silk. Some spiders will eat their used webs. If a spider is descending on a strand of silk, and needs to go back up on that strand, it will consume the strand on the way up!

Traditional Knowledge

From Taloyoak: Ellen Ittunga interviewing Mary Ittunga, July 2006

Spiders were either placed or rubbed onto newborn boys and probably girls, too. It was thought that the child would grow up to be swift or able to catch game every time they went out hunting. The practice of placing or rubbing the spider on the baby's hand [was] called aannguat.

Order: *Araneae*

Wolf Spiders

Lycosidae

Order: Araneae – True Spiders; Family: Lycosidae – Wolf Spiders

Members of the Lycosidae family are commonly known as wolf spiders in English. No specific Inuktitut names were found for this family during the research for this book. Please see the Araneae account for general names for spiders. Also, if you know of names for any species in this family, please contact the Nunavut Teaching and Learning Centre.

Description

Wolf spiders are large and dark with well-developed eyesight. These spiders are between three and thirty millimetres long. Both the **cephalothorax** and the **abdomen** on the wolf spider are usually as long as they are wide, meaning that these body sections are quite round. The **carapace** sometimes has dark bands. On the cephalothorax, the wolf spider has eight eyes divided into three rows. The first row has four small eyes in a straight line. Behind that are two larger eyes that look forward, and a little further back still are two more large eyes that look up. The spider has eight long legs attached to the cephalothorax. Each leg has three microscopic claws at the tip. Female wolf spiders are often larger than the males.

Life Cycle

Incomplete Metamorphosis:
These solitary spiders only come together to mate. To attract a female, the male spider waves his **pedipalps** and raises his front legs. Once the pair has mated, the female lays several dozen eggs and then spins a round egg sac in which to carry them. The sac remains attached to her **spinnerets**. When the **spiderlings** hatch, they are carried around on their mother's abdomen, clinging to her **setae** (specialized hairs), until they **moult** for the first time. During the time the spiderlings spend on their mother's body, they do not eat. Female wolf spiders exhibit considerable care for their eggs and young, which is not typical spider behaviour. The male spiders can live up to one year, while the females may live for as long as three years.

Photo by Carolyn Mallory

Food and Feeding

Wolf spiders mostly hunt insects for food, although large females have been known to eat small amphibians and reptiles in more southern locations. They hunt on the ground.

Habitat

Their dark, blotchy colouring provides wolf spiders with excellent camouflage, as they hide among dead leaves and debris. Except for one genus, the spiders in this family do not spin webs. Some of them dig holes for shelter and some of them hide under rocks, but many of them have no hiding places.

Range

Just over twenty species of wolf spiders have been found in the Arctic so far. A couple of those species were found in the High Arctic.

Did You Know?

Wolf spiders take their scientific name from the Greek word for wolf (*lycosa*) because of the way they ambush or run down their prey while hunting them.

Photo by Susan Aiken

Crab Spiders

Philodromidae

Order: Araneae – True Spiders; Family: Philodromidae – Crab Spiders

Members of the Philodromidae family are commonly known as crab spiders in English. No specific Inuktitut names were found for this family during the research for this book. Please see the Araneae account for general names for spiders. Also, if you know of names for any species in this family, please contact the Nunavut Teaching and Learning Centre.

Description

The crab spider has a flat **cephalothorax** which is longer than it is wide. It has two rows of four eyes. They can be various shades of yellow, off-white, or orange, and all have pale bands. When they are mature, they are between two and eight millimetres long. The **abdomen**—what we think of as the body of a spider—is a long oval, and is widest in the middle. They have **scale**-like hairs or bristles that lie down flat on their bodies. Their eight legs are long and skinny, and are almost all the same length, with two claws on the first legs. They have stiff, bristle-like hairs on their legs, as well. These, along with their claws, help them crawl around on slippery and sloping leaves.

Photo by Jorgen Lissner

To move around, crab spiders shuffle quickly and randomly, staying low to the ground. They are able to move like that because they have flexible, flat bodies and their legs face forward like those of a crab.

Life Cycle

Incomplete Metamorphosis:
The nymphs are born in the summer and are almost mature by the time they **overwinter**. They mature and mate in the spring, and lay eggs in early summer, when the cycle starts again. As do many spiders, the females build silk **egg sacs**.

Food and Feeding

Crab spiders are hunting spiders, not web-building spiders. In other words, they capture their insect prey without the use of webs.

Habitat

Most crab spiders live in grass or trees, but in the Arctic, they live on the tundra.

Range

These spiders are found in the Low Arctic from Alaska to Greenland. The spiders in this family that have been collected in the Low Arctic belong mostly to the genus *Thanatus*.

Did You Know?

Spiders do not have lungs like ours with which to breathe. Instead, they have openings in their abdomens called **book lungs**, which they use to respire. Spiders can have one or two pairs of book lungs. If they only have one pair, then they also use their **tracheae** to help with breathing.

Photo by Bev Wigney

Order: *Araneae*

143

Sheet Web Spiders and Dwarf Spiders

Lyniphiidae

Order: Araneae – True Spiders; Family: Lyniphiidae – Sheet Web Spiders and Dwarf Spiders

These spiders are commonly known as sheet web spiders and dwarf spiders in English. No specific Inuktitut names were found for this family during the research for this book. Please see the Araneae account for general names for spiders. Also, if you know of names for any species in this family, please contact the Nunavut Teaching and Learning Centre.

Description

Adults:
The spiders in this family are tiny, ranging from 1.7 to 7.2 millimetres in length. They have eight eyes, generally grouped in two horizontal rows of four. Some species (in the subfamily Erigoninae) have conspicuous lobes, swellings, or turrets on their **cephalothoraces**, while others do not. Male **pedipalps** are much larger than the females' and are structurally more complex. They are used both for mating and for grabbing prey. Female pedipalps may or may not have claws. Sheet web and dwarf spiders have fangs with venom.

These spiders have short legs. They are mostly dark and shiny, and their abdomens can be patterned or plain.

Life Cycle

Incomplete Metamorphosis:
Female sheet web and dwarf spiders put their eggs in silk egg sacs, which may be attached to their webs and watched, or attached to blades of grass and left alone. Small spiders hatch from the eggs. These young spiders look very much like the adults, except that they are smaller. Their colouring may change as they age. As spiders grow into adulthood, they shed their skins or **exoskeletons** many times before they reach full size.

Food and Feeding

Some Lyniphiidae spiders weave sheet-like, silky webs in the grass or on the tundra. They then sit under the webs and wait for small insects to walk across them. Once the insects are caught in the webs, the spiders bite through the silk, grab their prey, and eat them.

Other members of this family are **predators**.

Habitat

Spiders in the Lyniphiidae family spend most of their time under their webs, which they construct on the tundra. They may also be spotted walking on the tundra; however, these spiders are very small and largely go unnoticed.

Range

About a dozen species have been identified in the Low Arctic and a single species in the High Arctic thus far. It is likely that many species are still left to be discovered.

Did You Know?

Spiders in the Lyniphiidae family use their webs as balloons to transport themselves from one place to another.

Photo by Stephan Sollfors

Order: *Araneae*

Mites and Ticks

Subclass: Acari – Mites and Ticks

The Acari subclass is made up of mites and ticks. No specific Inuktitut names were found for this subclass during the research for this book. If you know of names for any species in this subclass, please contact the Nunavut Teaching and Learning Centre.

Acari is the largest subclass of the class Arachnida. Note: No ticks have been found yet in Nunavut, so this account refers only to mites.

Description

Mites differ from other arachnids in that they do not have a division between their two body parts. They are very tiny creatures; some are even microscopic. Mites are usually smaller than ten millimetres in length. Like spiders, they have four pairs of legs when they are mature adults. Mites also have a pair of **pedipalps**, which are small, leg-like appendages that help the mites direct food to their mouths.

These tiny creatures may be **parasitic**, or they may live independently. The mites that live independently often travel on other living creatures (insects or mammals) without feeding on them. The term for this is **phoresy**. They travel from place to place by hanging on to another animal (like people riding horses)—what a way to travel!

Life Cycle

Incomplete Metamorphosis:
Most mites lay eggs, which then become six-legged larvae. The larvae go through one to three stages of nymphhood, during which each larva grows an additional pair of legs. Finally, the larvae become adults, ready to reproduce.

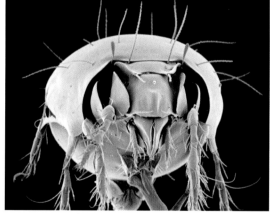

Photo by Valerie Behan-Pelletier

Food and Feeding

Mites are a very diverse group when it comes to feeding. Some mites feed on pollen, while others feed on the blood of mammals, such as lemmings. Still other mites feed on fungae, algae, or young bird chicks in their nests. Some tiny, soft-bodied mites feed on dead plants or leaves, while others feed on live plants. There are also mites that hunt insects or eat their eggs. And, of course, there are the parasitic mites that feed on their hosts, which may be insects, mammals, or birds. Mites sometimes consume micro-organisms in wet places and can be found scavenging dead animals.

Photo by Valerie Behan-Pelletier

Habitat

Mites have adapted to every type of habitat imaginable—they can survive extreme heat and cold. They live in saltwater, freshwater, and soil, and on other animals. More specifically, they can live in the tracheae of bees; in the ears of moths; in the hair follicles of humans; in the nasal cavities or feathers of ducks; and in the fabric of people's sofas, beds, and pillows. Sometimes they live just underneath human or dog skin. They can live under soil to a depth of ten metres or under water to a depth of five thousand metres.

Range

Mites are found in both the High and Low Arctic. At least one hundred species of Acari have been identified in the Arctic.

Did You Know?

Mites are almost inconceivably numerous and diverse. One square metre of boreal forest floor matter can contain up to one million mites. This number might be made up of two hundred species from fifty different families.

Other Interesting Mites

Subclass: Acari

Tetranychidae family – Spider Mites
Members of the Tetranychidae family are commonly known as spider mites in English.

These tiny creatures are usually between 0.3 and 0.8 millimetres long, with flat, oval-shaped bodies. At the ends of their **cephalothoraces**, spider mites have **chelicerae**, which enable them to feed on plant juices. Both males and females have **pedipalps**. The female's pedipalps resemble legs. The male's pedipalps are swollen at the tip and are used for transferring sperm to the female. Both male and female mites use their pedipalps to help them hold on to leaves while they feed. The female uses a silk-gland opening near her mouth to spin small webs.

Many species in this family feed on the leaves of plants, using their mouthparts to pierce and suck the juices out of plant leaves. It is assumed that most of the Arctic species feed on grasses.

Other Interesting Mites

Sarcoptes scabiei – Scabies
This small mite (0.2 to 0.4 millimetres) causes a contagious disease in humans called scabies. The mites dig a tunnel in the dead (top) layer of human skin, where they live for most of their lives. The female starts to lay eggs almost immediately once she has dug her tunnel, and it only takes the eggs ten to fourteen days to become adults that are ready to reproduce. Skin eventually reacts to these mites by becoming very itchy. Once diagnosed, this disease is treatable.

Dermatophagoides spp. – Dust Mites
Dust mites can be found in houses all over the world, although they prefer more humid locations. We are not able to see them with our eyes because they are so small, so we are not really aware of their presence unless we have allergies. These mites live in bedding, pillows, sofas, stuffed toys, and carpets. They eat dead human skin and other organic materials.

Ornithonyssus bursa – Bird Mites
As the name implies, bird mites are found on birds, and also in their nests. They are parasites that feed on bird blood. Bird nests provide a warm, safe place to live and, although their lifespans are short (seven days), bird mites can still produce an enormous number of new mites. Most of these mites will die after three weeks of not having any bird blood.

***Demodex* spp.** – Follicle Mites

Two types of follicle mites live on human beings. Members of the first species live in human hair follicles, while members of the second species live in the oil glands of the face (particularly around the nose, eyes, and forehead). Follicle mites occur on almost 100% of older people, but rarely cause health problems.

Did You Know?

There can be as many as twenty-five mites living in one hair follicle. A used mattress may contain from one hundred thousand to ten million dust mites.

Glossary

Abdomen

Abdomen: The posterior section of an insect's or spider's body.

Abdominal: Of or relating to the abdomen.

Apodous: Without feet or foot-like structures.

Aquatic: Living in water.

Arista

Arista: Threadlike attachment at the end of the antenna; usually a large bristle.

Arthropod: Animals without backbones, including insects, spiders, and crustaceans.

Autogenous: Able to make eggs without protein from a blood meal.

Book lungs: Sac-like arachnid respiratory organ with several folds arranged like the pages of a book.

Bristles: Thick hairs.

Calypter: Lobe at the base of a fly's wing.

Carapace: Protective covering on the backside of a spider's cephalothorax.

Bristles

Carnivore: Any animal that feeds on flesh.

Carnivorous: Feeding only on other animals.

Carrion: Decaying flesh of dead animals.

Cauda: Tail-like structure at the end of an aphid's abdomen.

Cells: Spaces between the veins in the wing.

Cephalothorax: The fused head and thorax of a spider.

Cells

Cerci (sing. cercus): Sensory organs, usually occurring in pairs at the end of the abdomen, that are sometimes used to detect the presence of nearby insects.

Chelicerae: Jaw-like appendages in arachnids.

Chrysalis (pl. chrysalids): Pupa of a butterfly.

Cerci

Clubbed: Enlarged at the tip (often with regards to antennae).

Cocoon: A silky case spun by larvae in order to protect themselves while they pupate.

Complete metamorphosis: The passage of an insect through four different stages of growth—egg, larva, pupa, and adult.

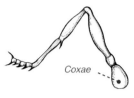
Compound eyes
Coxae

Compound eyes: Eyes that consist of many small visual units.

Convex: Domed, like an igloo.

Cornicle: Tubular projection from the abdomen of an aphid.

Coxa (pl. coxae): The first leg segment.

Detritus: Dead plant or animal matter.

Diapause: A long break during which an insect's body functions slow down and it waits for the right conditions under which to continue its life cycle.

Elytra

Egg sac: Specialized sac spun from silk in which the spider puts its eggs.

Elytra: Hard front wings.

Exoskeleton: Rigid external covering that provides an insect with support and protection.

Fusiform

External digestion: Process of injecting digestive juices into food and then sucking up the partially digested food in a liquid form.

Fusiform: Broad in the middle and tapered at each end.

Gall: A construction that occurs when plant cells are altered by a chemical injected by an insect, allowing the larva(e) (one or several) to live inside.

Gelatinous: Jelly-like.

Genitalia: Sexual organs.

Gills: Respiratory organs for breathing under water.

Grub
Halteres

Girdle: Circular silk string that surrounds the chrysalis and by which the chrysalis is suspended.

Grub: The thick, wormlike larvae of some beetles and other insects.

Halteres: Reduced hindwings of flies, used for balance when flying.

Hamuli: Tiny hooks that attach the hymenopterans' front wings to their hindwings when in flight.

Haustellum (pl. haustella): A long, tongue-like sucker.

Hemoglobin: The iron-containing pigment in red blood cells.

Herbivore: An animal that only eats plants.

Herbivorous: Feeding only on plants.

Honeydew: Sugary faeces produced by some insects, most typically aphids.

Host: Animal or plant on which a parasite lives.

Incomplete metamorphosis: The passage of an insect through three different stages of growth—egg, larva, and adult—but not through a pupal stage.

Instar: Period of growth followed by a change of skin.

Invertebrate: An animal without a backbone.

Iridescent: With rainbow-like colours that shift and change in the light.

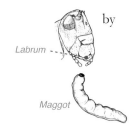

Labium (pl. labia): Lower lip.

Labrum (pl. labra): Upper lip.

Maggot: Common name for most fly larva.

Mandibles: Jaws.

Marine: Living by or in the sea.

Maxillae: Paired mouthpart structure beside the jaw.

Membranous: Thin, flexible, and almost transparent.

Micro-organisms: Tiny critters that cannot be seen with the naked eye.

Moult: Process by which an arthropod sheds its skin and grows a new one.

Myiasis: Infestation of the body by larvae of flies (maggots).

Nocturnal: Active at night.

Nymph: Immature forms of insects that undergo incomplete metamorphosis; also the immature forms of spiders and mites.

Obtect: (of an insect pupa or chrysalis) Covered in a firm case, with legs, wings, and antennae glued to the body.

Ocelli: Simple eyes.

Omnivore: Organism that eats both plants and animals.

Omnivorous: Feeding on both plants and animals.

Oothecae: The specialized egg cases of cockroaches and praying mantids.

Overwinter: Live through the winter.

Oviduct: Tube through which an egg passes from the ovary.

Ovipositors: Tubes through which eggs are deposited.

Ovisac: Sac in which eggs are kept until the larvae are almost ready to come out.

Palps: Feeler-like parts of the mouth.

Papillae (sing. papilla): Little protrusions on the body.

Parasite: An organism that lives on a host organism and contributes nothing to the host.

Parasitoids: Organisms that live and feed

on host organisms until the hosts are eventually killed.

Parthenogenesis: Reproduction without fertilization by a male.

Pedipalps: Feeler-like parts of an arachnid's mouth.

Phoresy (adj. phoretic): An organism's

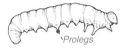

(e.g., a mite's) practice of travelling on the body of another, without being a parasite.

Pollinators: Creatures (mostly insects and birds) that help transfer plant pollen from the male part to the female part so that reproduction can take place.

Predaceous: Hunting and eating other animals.

Proboscis: Extended, beak-like mouthparts.

Prolegs: Appendages that are not quite legs.

Pronotum (pl. pronota): Hardened cover of the first segment of the thorax.

Pupa (pl. pupae): An insect in its immature form between larva and adult.

Pupal stage: Period of rest during which the larva transforms into an adult and develops wings.

Puparium: A case that is formed by the hardening of the final larval skin.

Pupate: Become a pupa.

Queen: The female in a colony of bees or ants whose job it is to lay all of the eggs.

Recurrent vein: A vein that turns back, reversing its direction.

Rostrum (pl. rostra): Snout-like extension of the mouth.

Scales: Small, overlapping protective coverings.

Scavenger: Animal that feeds on carrion, dead plant material, and garbage.

Sclerites: Hardened plates.

Sclerotized: Hardened.

Scutellum (pl. scutella): Hard, triangular segment of the thorax that acts as a shield.

Scutum (pl. scuta): Hard part of the thorax in front of the scutellum.

Setae: Specialized hair.

Social parasite: A parasite that chooses animals of its same type as hosts (i.e., one bee species living on another).

Spermatheca: (in a female invertebrate) A receptacle in which sperm is stored after mating.

Spiderlings: Young spiders.

Spinnerets: Organ through which silk is produced.

Spiracles: External respiratory openings.

Stylets: Needle-like structures that form part of some piercing mouthparts.

Subimago: A stage in the development of some insects (such as mayflies) between the nymph and the imago.

Subscutellum (pl. subscutella): Ridge or lobe under the scutellum.

Supercool: (of a living organism) Cool the body to an extreme degree in order to survive temperatures below the freezing point of water.

Syphon: A tubular organ with which air is taken in or expelled.

Tarsus (pl. tarsi): Segment at the end of the leg, somewhat similar to a foot.

Terrestrial: Living on land.

Thorax: Middle body section between the head and the abdomen.

Tibia: The fourth segment of the leg.

Trachea (pl. tracheae): Breathing tubes that transmit oxygen throughout the body.

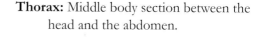

Tubercles: Projections or outgrowths.

Tympanum (pl. tympana): Vibrating membrane similar to an eardrum.

Urogomphi: Paired, tail-like appendages on beetle larvae.

Vertebrates: Animals with backbones.

Wing buds: The beginning of wings in hemipterans, which grow larger with each moult until they become wings.

Wing margin: Dark, thickened edge of the wing.

Wingpads: The partially developed wings of insect nymphs that have undergone incomplete metamorphosis.

Bibliography

Anderson, Robert S. and Stewart B. Peck. *The Insects and Arachnids of Canada, Part 13: the Carrion Beetles of Canada and Alaska: Coleoptera: Silphidae and Agyrtidae*. Ottawa: Agriculture Canada, 1985.

Bickel, D. J. and C. E. Dyte. "Dolichopodidae." Hawaii Biological Survey. http://hbs.bishopmuseum.org/aocat/doli.html

Borror, Donald J. and Richard E. White. *A Field Guide to Insects of America North of Mexico*. Boston: Houghton Mifflin Company, 1970.

Borror, Donald J., Dwight M. De Long, and Charles A. Triplehorn. *An Introduction to the Study of Insects*, 5th ed. Philadelphia: Saunders College Publishing, 1981.

Bousquet, Yves. "Ground Beetles (Coleoptera: Carabidae)." Research Branch, Agriculture Canada. http://naturewatch.ca/mixedwood/beetles/intro.htm

Buck, M., S. A. Marshall, and D. K. B. Cheung. "Identification Atlas of the Vespidae (Hymenoptera, Aculeata) of the Northeastern Nearctic Region." *Canadian Journal of Arthropod Identification* 5 (19 February 2008): 1–492.

Canadian Biodiversity Information Facility. "Ground Beetles of Canada." Government of Canada. http://www.cbif.gc.ca/spp_pages/carabids/phps/a_e.php.

Canadian Biodiversity Information Facility. "Family Nymphalidae." Government of Canada. http://www.cbif.gc.ca/spp_pages/butterflies/families/nymphalidae_e.php.

Danks, Hugh V., Olag Kukal, and R. A. Ring. "Insect Cold-Hardiness: Insights from the Arctic." *Arctic* 47, no. 4 (1994): 391–404.

Danks, Hugh V. "Seasonal Adaptations in Arctic Insects." *Integrative and Comparative Biology* 44, no. 2 (2004): 85–94.

Dondale, Charles D. and James H. Redner. *The Insects and Arachnids of Canada, Part 5: The Crab Spiders of Canada and Alaska: Aramae: Philodromidae and Thomisidae*. Ottawa: Agriculture Canada, 1978.

Downes, J. A. "Arctic Insects and Their Environment." *The Canadian Entomologist* 96 (1964): 279–307.

Downes, J. A. "Adaptations of Insects in the Arctic." *Annual Reviews in Entomology* 10 (1965): 257–274.

Gagné, Raymond J. "Cecidomyiidae." Hawaii Biological Survey. http://hbs.bishopmuseum.org/aocat/cecido.html.

Gullan, P. J. and P. S. Cranston. *The Insects: An Outline of Entomology*, 3rd ed. Malden: Blackwell Publishing Ltd., 2005.

Harington, C. R., ed. *Canada's Missing Dimension: Science and History in the Canadian Arctic Islands*. Ottawa: Canadian Museum of Nature, 1990.

Kevan, Peter G. "Thermoregulation in Arctic Insects and Flowers: Adaptation and Co-adaptation in Behaviour, Anatomy, and Physiology." In *Thermal Physiology*, edited by J. B. Mercer, 747–753. Amsterdam: Excerpta Medica, 1989.

Kurahshi, Hiromu. "Calliphoridae." Hawaii Biological Survey. http://hbs.bishopmuseum.org/aocat/calliphoridae.html

Layberry, Ross A., Peter W. Hall, and J. Donald Lafontaine. *The Butterflies of Canada*. Toronto: University of Toronto Press, 1998.

Lyon, William F. "Midges and Crane Flies." Ohio State University. http://ohioline.osu.edu/hyg-fact/2000/2129.html.

Maddison, David R. "Carabidae: Ground Beetles and Tiger Beetles." The Tree of Life Web Project. http://tolweb.org/Carabidae/8895/2006.04.11

Marshall, Stephen A. *Insects: Their Natural History and Diversity: With a Photographic Guide to Insects of Eastern North America*. Richmond Hill, ON: Firefly Books, 2006.

Maw, H. E. L., R. G. Foottit, K. G. A. Hamilton, and G. G. E. Scudder. *Checklist of the Hemiptera of Canada and Alaska*. Ottawa: NRC Research Press, 2000.

Mead, F. W. and T. R. Fasulo. "Darkwinged Fungus Gnats, Bradysia spp. (Insecta: Diptera: Sciaridae)." University of Florida Institute of Food and Agricultural Sciences. http://edis.ifas.ufl.edu/in372.

Milne, Lorus and Margery Milne. *National Audubon Society Field Guide to North American Insects and Spiders*. New York: Alfred A. Knopf, 2004.

Moore, Marianne V. and Richard E. Lee, Jr. "Surviving the Big Chill: Overwintering Strategies of Aquatic and Terrestrial Insects." *American Entomologist* 37 (1991): 111–118.

Myers, Phil. "Insecta." Museum of Zoology, University of Michigan. http://animaldiversity.ummz.umich.edu/site/accounts/information/Insecta.html

Nelson, Robert E. "Bioclimatic Implications and Distribution Patterns of the Modern Ground Beetle Fauna (Insecta: Coleoptera: Carabidae) of the Arctic Slope of Alaska, U.S.A." *Arctic* 54, no. 4 (December 2001): 425–430.

O'Toole, Christopher, ed. *Firefly Encyclopedia of Insects and Spiders*. Richmond Hill, ON: Firefly Books, 2002.

Ring, Richard A. and D. Tesar. "Adaptations to Cold in Canadian Arctic Insects." *Cryobiology* 18, no. 2 (April 1981): 199–211.

Schmid, F. *The Insects and Arachnids of Canada, Part 7: Genera of the Trichoptera of Canada and Adjoining or Adjacent United States*. Ottawa: NRC Research Press, 1998.

Vickery, Vernon R. and D. Keith McE. Kevan. *The Insects and Arachnids of Canada, Part 14: The Grasshoppers, Crickets, and Related Insects of Canada and Adjacent Regions: Ulonata: Dermaptera, Cheleutoptera, Notoptera, Dictuoptera, Grylloptera, and Orthoptera*. Ottawa: Agriculture Canada, 1985.

Whitfield, James B., Won-Young Choi, Alejandro A. Valerio, Josephine Rodriguez, and Andrew R. Deans. "Braconidae." The Tree of Life Web Project. http://tolweb.org/Braconidae/23447/2004.06.10

Wood, D. M., P. T. Dang, and R. A. Ellis. *The Insects and Arachnids of Canada, Part 6: The Mosquitoes of Canada: Diptera: Culicidae*. Ottawa: Agriculture Canada, 1979.

Index

A

aabjiq. See Hymenoptera: social wasps
aanngiq. See Hymenoptera: social wasps
aasivak. See Araneae: true spiders
abdomen (*general*), 8, 10
    abdominal prolegs, 76
    of aphids, 34
    of arachnids, 136
    of balloon flies, 78
    of bees, 130
    of black flies, 66–67
    of blowflies, 90
    of butterflies and moths (*general*), 96–97
    of caddis flies, 116
    of carrion beetles, 46
    of cockroaches, 24
    of common sawflies, 122
    of crab spiders, 142
    of crane flies, 60
    of dung flies, 84
    of flies (*general*), 56
    of houseflies, 88
    of ichneumon wasps, 126
    of ladybugs, 52
    of long-legged flies, 80
    of looper moths, 110
    of mayflies, 20
    of mosquitoes, 62
    of no-see-ums, 68
    of predaceous diving beetles, 40
    of root-maggot flies, 86
    of rove beetles, 48–49
    of sheet web and dwarf spiders, 144
    of social wasps, 128
    of stoneflies, 22
    of tachinid flies, 94
    of true spiders, 138
    of warble and botflies, 92
    of wolf spiders, 140
Acari. *See* Araneae: mites and ticks (*general*); spider mites; scabies; dust mites; bird mites; follicle mites
Aedes. See Diptera: mosquitoes

Agriades glandon, 105
anangiq. *See* Diptera: blackflies
anangiqpa. *See* Diptera: houseflies
anangirjuaq. *See* Diptera: houseflies
ananngiq. *See* Diptera
Anthomyiidae. *See* Diptera: root-maggot flies
Apaulina sapphira, 90–91
Aphididae. *See* Hemiptera: aphids
Apidae (*general*), 3
    aphids and, 35
    bumblebees, 130–133
    hoverflies and, 82–83
    honeybees, 130
    subfamily Apinae, 130–133
    warble and botflies and, 92
Apinae. *See* Apidae
Apocrita, 118–120, 124–133
aqammukitaaq. *See* Coleoptera: predaceous diving beetles
aquatic lifestyles
    of flies (*general*), 56–57
    of hoverflies, 82
    of mayflies, 20–21
    of mosquitoes, 62–63
    of non-biting midges, 71
    of no-see-ums, 69
    of predaceous diving beetles, 40–41
    of stoneflies, 22–23
    of true bugs and homopteran insects (*general*), 31–32
    of water scavenger beetles, 44–45
Arachnida (*general*), 136–137, 146
Araneae
    bird mites, 149
    crab spiders, 142–143
    dust mites, 148
    follicle mites, 149
    mites and ticks (general), 146–147
    sheet web and dwarf spiders, 144–145
    scabies, 148
    spider mites, 148
    true spiders, 138–139
    wolf spiders, 140–141
Auchenorrhyncha. *See* Hemiptera
arista, 90, 94
arlungajuq. *See* Diptera: warble and botflies
Arthropoda, 4, 8, 10, 23, 94, 129
aullak. *See* Coleoptera: rove beetles

autviq. See Lepidoptera (*general*)

B
balloon flies. *See* Diptera
bees. *See* Apidae
beetles. *See* Coleoptera
bigligiak. See Coleoptera: click beetles
bird mites. *See* under Araneae
Blattodea, 24–25
Bombus hyperboreus, 131
Bombus polaris, 131
book lungs, 136, 143
Boreallus atricups, 91
Braconidae. *See* Hymenoptera: parasitic wasps
bristles, 80, 84, 88, 90, 94, 142

C
caddis flies. *See* Trichoptera
calypters, 58, 84, 86, 90, 94
carapaces, 136, 138, 140
carnivores, 40, 69, 77
carrion, 46–47, 57, 89, 91
cauda, 34–35
Cecidomyiidae. *See* Diptera: gall midges
cells, 10, 13, 74, 124
Ceratopogonidae. *See* Diptera: no-see-ums
cerci, 20, 22, 24
chelicerae, 136, 138, 148
Chironomidae. *See* Diptera: non-biting midges
chrysalids, 96, 103, 105
clubbed antennae, 44, 46, 48, 52, 54, 98, 102, 106
cocoons
    bees and, 131
    black flies and, 66–67
    braconid wasps and, 125
    butterflies and moths (*general*) and, 97, 112–113, 115
    caddis flies and, 117
    long-legged flies and, 81
    non-biting midges and, 70
    rove beetles and, 48
    sawflies and, 122
cockroaches. *See* Blattodea
Coleoptera (*general*), 10, 24
    carrion beetles, 46–47
    click beetles, 50–51

    ground beetles, 38–39
    ladybugs, 52–53
    predaceous diving beetles, 40–43
    rove beetles, 48–49
    water scavenger beetles, 44–45
    weevils, 54–55
Colias species, 103
common sawfly. *See* Tenthredinidae
complete metamorphosis (*general*), 11–12
    balloon flies and, 78
    bees, wasps, ants, and sawflies (*general*), 118–119, 131
    beetles (*general*) and, 36
    blowflies and, 90
    braconid wasps and, 124–125
    brush-footed butterflies and, 106–107
    butterflies and moths (*general*) and, 97
    caddis flies and, 116–117
    carrion beetles and, 46
    click beetles and, 50
    common sawflies and, 122
    crane flies and, 61
    dark-winged fungus gnats and, 72
    dung flies and, 84–85
    flies (*general*) and, 57
    gall midges and, 74
    gossamer-winged butterflies and, 104–105
    ground beetles and, 38
    horseflies and, 77
    houseflies and, 88
    hoverflies and, 82–83
    leafroller moths and, 100
    long-legged flies and, 80–81
    looper moths and, 111
    mosquitoes and, 63
    non-biting midges and, 70
    no-see-ums and, 68
    owlet moths and, 115
    parasitic wasps and, 126–127
    predaceous diving beetles and, 40
    root-maggot flies and, 86–87
    rove beetles and, 48
    snout moths and, 108
    social wasps and, 128–129
    tachinid flies and, 94–95
    tussock moths and, 112–113
    warble and botflies and, 93

water scavenger beetles and, 44
white and sulphur butterflies and, 102–103
winter crane flies and, 58
compound eyes (*general*), 8–9
bees and, 130
beetles (*general*) and, 36
black flies and, 66
butterflies and moths (*general*) and, 96
caddis flies and, 116
cockroaches and, 24
crane flies and, 60
dark-winged fungus gnats and, 72
dung flies and, 84
flies (*general*) and, 56
gall midges and, 74
ground beetles and, 38
horseflies and, 76
houseflies and, 88
leafroller moths and, 100
mosquitoes and, 62
non-biting midges and, 70
owlet moths and, 114
tachinid flies and, 94
true bugs and homopteran insects (*general*) and, 30
convex shape, 44, 52
Culicidae. *See* Diptera: mosquitoes
Culiseta genus, 63
Curculionidae. *See* Coleoptera: weevils

D

Demodex spp. *See* Araneae: follicle mites
Dephacidae family, 32
Dermatophagoides spp. *See* Araneae: dust mites
detritus, 15, 23
diapause, 14, 105
Diptera (*general*), 56–57
balloon flies, 78–79
blackflies, 66–67
blowflies, 90–91
caribou warble fly, 93
crane flies, 59–60
dance flies, 78–79
dark-winged fungus gnats, 72–73
deer flies, 76
dung flies, 84–85

    gall midges, 74–75
    horseflies, 76–77
    houseflies, 88–89
    hoverflies, 82–83
    long-legged flies, 80–81
    mosquitoes, 62–65
    non-biting midges, 70–71
    no-see-ums, 68–69
    root-maggot flies, 86–87
    tachinid flies, 94–95
    warble and botflies, 92–93
    winter crane flies, 58–59
Dolichopodidae. *See* Diptera: long-legged flies
Dolichovespula norwegica, 129
dung flies. *See* Diptera: dung flies
dust mites. *See* Araneae: dust mites
dwarf spiders. *See* Araneae: dwarf spiders
Dytiscidae. *See* Coleoptera: predaceous diving beetles

E

egg sacs, 126, 138, 140, 142, 144
egotak. *See* Diptera: warble flies
Elateridae. *See* Coleoptera: click beetles
elytra (*general*), 10
    of beetles (*general*), 36
    of carrion beetles, 46
    of ground beetles, 38
    of ladybugs, 52
    of predaceous diving beetles, 40–41
    of rove beetles, 48
Empididae. *See* Diptera: balloon flies
Ephemeroptera, 20–21
exoskeleton (*general*), 8, 10
    of beetles (*general*), 36
    of click beetles, 50
    of sheet web and dwarf spiders, 144
external digestion, 39

F

flies. *See* Diptera
follicle mites. *See* Araneae: follicle mites
front wings (*general*), 10
    of bees, 130
    of beetles (*general*), 36
    of butterflies and moths (*general*), 96

    of caddis flies, 116
    of carrion beetles, 46
    of cockroaches, 24
    of ground beetles, 38
    of hymenopterans (*general*), 118
    of ladybugs, 52
    of leafroller moths, 100
    of looper moths, 110
    of mayflies, 20
    of owlet moths, 114
    of predaceous diving beetles, 40
    of rove beetles, 48
    of snout moths, 108
    of true bugs and homopteran insects (*general*), 30–31
fusiform shape, 76

G

gall midges. *See under* Diptera
galls, 74–75, 123, 125, 127
genitalia, 78, 80
Geometridae. *See* Lepidoptera: looper moths
gills
    of blackflies, 66
    of caddis flies, 116–117
    of mayflies, 20
    of stoneflies, 22
    of winter crane flies, 58
girdle, 105
gossamer-winged butterflies. *See under* Lepidoptera
ground beetles. *See under* Coleoptera
grubs
    beetles (*general*), 36
    bumblebees, 130
    carrion beetles, 46
    parasitic wasps, 126
    social wasps, 128
    weevils, 54
Gynaephora groenlandica, 113

H

hakalikitaak. *See* Lepidoptera
halteres, 10, 56, 60
hamuli, 118
haqalikitaaq. *See* Lepidoptera
haustellum, 96

Hemiptera (*general*), 30–33
    and bioluminescence, 51
    aphids, 34–35
hemoglobin, 70
herbivores, 37, 54–55, 69, 119
Heteroptera (*general*), 30–33
hindwings, 10
    of beetles (*general*), 36
    of bees, 130
    of butterflies and moths (*general*), 96
    of caddis flies, 116
    of cockroaches, 24
    of crane flies, 60
    of flies, 56
    of hymenopterans (*general*), 118
    of ladybugs, 52
    of mayflies, 20
    of owlet moths, 114
    of rove beetles, 48
    of snout moths, 108
    of stoneflies, 22
    of true bugs and homopteran insects (*general*), 30–31
Homoptera (*general*), 30–33
honeydew
    aphids and, 34–35
    blowflies and, 91
    braconid wasps and, 125
    gossamer-winged butterflies and, 104–105
    hymenopterans (*general*) and, 119
    tachinid flies and, 95
horseflies. *See under* Diptera
hosts
    braconid wasps and, 124–125
    flies (*general*) and, 57
    houseflies and, 89
    hymenopterans (*general*) and, 119
    lice and, 26–27
    looper moths and, 111
    mites (*general*) and, 147
    owlet moths and, 115
    parasitic wasps and, 126–127
    tachinid flies and, 94–95
    warble and botflies and, 93
houseflies. *See under* Diptera
hoverflies. *See under* Diptera
Hydrophilidae. *See* Coleoptera: water scavenger beetles

Hymenoptera (*general*), 118–121
    bees, 130–133
    braconid wasps, 124–125
    common sawflies, 122–123
    parasitic wasps, 126–127
    social wasps, 128–129
Hydroderma tarandi. See Diptera: caribou warble fly

I

Ichneumonidae. *See* Hymenoptera: parasitic wasps
igupsaq. *See* Hymenoptera: bees
iguptaq. *See* Hymenoptera: bees
iguptaujaq. *See* Hymenoptera: social wasps
igutsaq. *See* Diptera: warble flies; Hymenoptera: bees
iguttaq. *See* Diptera: warble flies; Hymenoptera: social wasps, bees
ikarmikitaa. *See* Coleoptera: predaceous diving beetles, water scavenger beetles
imarmiutait. *See* Coleoptera: predaceous diving beetles, water scavenger beetles
incomplete metamorphosis (*general*), 11–12
    aphids and, 34–35
    arachnids (*general*) and, 136
    cockroaches and, 25
    crab spiders and, 142
    lice and, 27
    mayflies and, 20–21
    mites and ticks (*general*) and, 146
    sheet web and dwarf spiders and, 144
    stoneflies and, 22–23
    true bugs and homopteran insects (*general*) and, 32
    true spiders and, 138
    wolf spiders and, 140
instars (*general*), 11–12
    bees (*general*) and, 131
    beetles (*general*) and, 36
    black flies and, 67
    blowflies and, 91
    butterflies and moths (*general*) and, 97
    caddis flies and, 117
    carrion beetles and, 46
    common sawflies and, 122
    dark-winged fungus gnats and, 73
    dung flies and, 84–85
    flies (*general*) and, 56
    ground beetles and, 38
    horseflies and, 77
    houseflies and, 88
    hoverflies and, 83
    looper moths and, 111
    long-legged flies and, 81
    midges and, 70
    mosquitoes and, 63
    no-see-ums and, 68
    predaceous diving beetles and, 41
    root-maggot flies and, 86
    tachinid flies and, 94
    warble and botflies and, 93
    white and sulphur butterflies and, 102–103
    winter crane flies and, 58
invertebrates, 4, 85
iqqamukisaat. *See* Coleoptera: predaceous diving beetles, water scavenger beetles
iqqaq. *See* Coleoptera: predaceous diving beetles, water scavenger beetles
iqqiq. *See* lice
iridescence, 76

K

kaligoalik (var. *kiktogiak*). *See* Diptera: mosquitoes
kikturiaqsiuyuyuq. *See* Diptera: horseflies
kingok. *See* Coleoptera: predaceous diving beetles, water scavenger beetles
kittuqsaq. *See* Hymenoptera: braconid wasps; Hymenoptera: parasitic wasps
kittusalik. *See* Diptera: non-biting midges
koglogiak. *See* Lepidoptera (*general*)

kumak (var. *kumaru, kumaruq*). *See* Hemiptera: aphids
kumaq. *See* lice
kumaujaq. *See* Diptera: blackflies

L

labium, 9, 118
labrum, 9, 50, 54
ladybugs. *See under* Coleoptera
leafroller moths. *See under* Lepidoptera
Lepidoptera (*general*), 96–99
    brush-footed butterflies, 106–107
    gossamer-winged butterflies, 104–105
    leafroller moths, 100–101
    looper moths, 110–111
    owlet moths, 114–115
    snout moths, 108–109
    tussock moths, 112–113
    white and sulphur butterflies, 102–103
lice, 26–28
long-legged flies. *See under* Diptera
looper moths. *See under* Lepidoptera
lungs, 58, 136, 143
Lycaenidae. *See* Lepidoptera: gossamer-winged butterflies
Lycosidae. *See* Araneae: wolf spiders
Lymantridae. *See* Lepidoptera: tussock moths
Lyniphiidae. *See* Araneae: sheet web and dwarf spiders

M

maggots
    blowflies, 90–91
    braconid wasps, 124
    flies (*general*), 56–57
    gall midges, 74
    hoverflies, 82–83
    long-legged flies, 80
    root-maggot flies, 86–87
    tachinid flies, 94
mandibles (*general*), 9
    of beetles (*general*), 36
    of caddis flies, 116–117
    of click beetles, 50
    of dung flies, 84
    of ground beetles, 38
    of horseflies, 76
    of houseflies, 88
    of predaceous diving beetles, 40–41
    of rove beetles, 48
    of stoneflies, 23
    of tachinid flies, 94
    of warble and botflies, 92
    of water scavenger beetles, 44–45
    of weevils, 54
maxillae, 9, 118, 130
mayflies. *See* Ephemeroptera
micro-organisms, 71, 89, 147
milogiak (var. *milogiakyoak*). *See* Diptera: dung flies
milookatak. *See* Diptera: houseflies
milugialaaq. *See* Diptera: houseflies
milugiaq. *See* Diptera: warble flies
minguq. *See* Coleoptera (*general*)
miqquligiaq (var. *miquligia*). *See* Lepidoptera
mites. *See* Araneae: mites and ticks (*general*)
mosquitoes. *See under* Diptera
moths. *See* Lepidoptera
moulting (*general*), 11–12
    bees, wasps, ants, and sawflies (*general*) and, 118–119
    beetles (*general*) and, 36
    braconid wasps and, 124
    brush-footed butterflies and, 107
    butterflies and moths (*general*) and, 97
    carrion beetles and, 46
    cockroaches and, 25
    crane flies and, 61
    ladybugs and, 52
    leafroller moths and, 100
    mayflies and, 20–21
    non-biting midges and, 70

owlet moths and, 115
parasitic wasps and, 127
predaceous diving beetles and, 41–42
snout moths and, 108
stoneflies and, 23
true spiders and, 138
tussock moths and, 112
water scavenger beetles and, 44
white and sulphur butterflies and, 102
wolf spiders and, 140
Muscidae. *See* Diptera: houseflies

N

nanuujaq. *See* Araneae: true spiders
Nematocera, 56. *See also* Diptera (*general*)
niaqusiugjujuq. *See* Diptera: dung flies
nigjuarjak (var. *nigjuk*). *See* Araneae: true spiders
niviaqyuk (var. *niviarjuit*). *See* Diptera: non-biting midges
nivik. *See* Diptera: non-biting midges
nivingasuuq. *See* Araneae: true spiders
niviuvait. *See* Diptera: non-biting midges
niviuvak (var. *niviovak*). *See* Diptera
Noctuidae. *See* Lepidoptera: owlet moths
non-biting midges. *See under* Diptera
no-see-ums. *See under* Diptera
Nymphalidae. *See* Lepidoptera: brush-footed butterflies

O

ocelli
of bees, wasps, ants, and sawflies (*general*), 118
of bumblebees, 130
of butterflies and moths (*general*), 97
of caddis flies, 116
of common sawflies, 122
of dark-winged fungus gnats, 72
of leafroller moths, 100
of owlet moths, 114
of true bugs and homopterans, 30
of true spiders, 138
of tussock moths, 112

of winter crane flies, 58
Oestridae. *See* Diptera: warble and botflies
oothecae, 25
overwintering (*general*), 14
aphids and, 34–35
blackflies and, 67
blowflies and, 91
bumblebees and, 131
butterflies and moths (*general*) and, 97
common sawflies and, 122
crab spiders and, 142
crane flies and, 61
gall midges and, 74
gossamer-winged butterflies and, 104
ground beetles and, 39
horseflies and, 77
hoverflies and, 83
ladybugs and, 52–53
leafroller moths and, 100
looper moths and, 111
mosquitoes and, 63
non-biting midges and, 70
owlet moths and, 115
predaceous diving beetles and, 40
root-maggot flies and, 86
social wasps and, 128–129
weevils and, 55
white and sulphur butterflies and, 103
oviducts, 132
ovipositors, 118, 122, 124–127
ovisacs, 95
owlet moths. *See under* Lepidoptera

P

palps
of balloon flies, 78
of bumblebees, 130
of butterflies and moths (*general*), 96
of caddis flies, 116
of dung flies, 84
of flies (*general*), 56
of long-legged flies, 80
of owlet moths, 114
of root-maggot flies, 86
of snout moths, 108

of water scavenger beetles, 44
of weevils, 54
pedipalps, 136, 138, 140, 144, 146, 148
papillae, 90
parasites (*general*), 12
    aphids, 34
    beetles (*general*), 37
    blowflies, 91
    braconid wasps, 124–125
    bumblebees, 131
    ecto-parasites, 26
    flies (*general*), 57
    houseflies, 89
    hoverflies, 83
    lice, 26–27
    mites and ticks (*general*), 146–149
    non-biting midges, 71
    parasitic wasps, 126–127
    parasitoid wasps (*general*), 119–120
    rove beetles, 49
    social wasps, 129
    tachinid flies, 94–95
    tussock moths and, 113
    warble and botflies, 92–93
parthenogenesis, 14–15, 32, 34
Pediculus humanus capitis. *See* lice
Pediculus humanus corporis. *See* lice
Philodromidae. *See* Araneae: crab spiders
Phthiraptera. *See* lice
Phthirus pubis. *See* lice
Pieridae. *See* Lepidoptera: white and sulphur butterflies
Plecoptera, 22–23
pollinators, 69
pootoogooqsiut. *See* Coleoptera: predaceous diving beetles, water scavenger beetles
predators
    ants protecting aphids from, 35
    ants protecting gossamer-winged butterflies from, 105
    arachnids (*general*), 137
    bees, wasps, ants, and sawflies (*general*), 119
    balloon flies, 79

    beetles (*general*), 37
    caddis flies, 117
    click beetles, 51
    crane flies, 61
    flies (*general*), 56–57
    gall midges, 75
    ground beetles, 39
    hoverflies, 83
    looper moths, 111
    long-legged flies, 81
    non-biting midges, 71
    predaceous diving beetles, 41
    rove beetles, 49
    sheet web and dwarf spiders, 144
    stoneflies, 23
    true spiders, 139
    tussock moths, 113
    water scavenger beetles, 45
predaceous diving beetles. *See under* Coleoptera
proboscises
    of blackflies, 66–67
    of blowflies, 90
    of brush-footed butterflies, 106
    of hoverflies, 82
    of leafroller moths, 100
    of long-legged flies, 80
    of mosquitoes, 62
    of non-biting midges, 70
    of no-see-ums, 68
    of owlet moths, 114
    of snout moths, 108
    of tussock moths, 112
prolegs
    of balloon flies, 78
    of bees, wasps, ants, and sawflies (*general*), 118
    of blackflies, 66
    of butterflies and moths (*general*), 97
    of caddis flies, 116
    of common sawflies, 122
    of horseflies, 76
    of houseflies, 88
    of looper moths, 110

of non-biting midges, 70
of owlet moths, 114
of white and sulphur butterflies, 102
pronota, 24, 46, 50, 52
Protophormia terranovae. See Diptera: blowflies
pupae (*general*), 12
    of balloon flies, 78
    of bees, wasps, ants, and sawflies (*general*), 119
    of beetles (*general*), 36
    of blackflies, 66–67
    of blowflies, 91
    of braconid wasps, 125
    of brush-footed butterflies, 107
    of bumblebees, 131
    of butterflies and moths (*general*), 97
    of caddis flies, 117
    of carrion beetles, 46–47
    of click beetles, 51
    of common sawflies, 122
    of crane flies, 61
    of dung flies, 85
    of flies (*general*), 57
    of gall midges, 74–75
    of gossamer-winged butterflies, 105
    of ground beetles, 38
    of horseflies, 77
    of houseflies, 88
    of hoverflies, 83
    of ladybugs, 52–53
    of leafroller moths, 100
    of looper moths, 111
    of long-legged flies, 80–81
    of mosquitoes, 62–63
    of no-see-ums, 68
    of non-biting midges, 70
    of owlet moths, 115
    of parasitic moths, 126–127
    of predaceous diving beetles, 41
    of root-maggot flies, 86
    of rove beetles, 48
    of snout moths, 108
    of social wasps, 128
    of tachinid flies, 95
    of true bugs and homopteran insects, 32
    of tussock moths, 112–113
    of warble flies, 93
    of water scavenger beetles, 44
    of weevils, 55
    of white and sulphur butterflies, 102–103
    of winter crane flies, 58
pupal stages. See instars
puparia, 85, 88

Q

qalirualik (var. *qalirulik, qaurulliq*). See Coleoptera: ground beetles
qauguklik. See Coleoptera: click beetles
qaumaja (var. *qaumajaq*). See Diptera: blackflies
qikturiaq. See Diptera: mosquitoes
qitirulliq. See Diptera
qitturiakalla. See Diptera: non-biting midges
quaraq. See Hemiptera: aphids
queens
    of bees, wasps, ants, and sawflies (*general*), 119
    of bumblebees, 130–132
    of social wasps, 128–129
quglugiaq (var. *qulluriaq*). See Lepidoptera (*general*)
qupirrualaat. See Hemiptera: aphids
quqsuqtuq (tarralikitaaq). See Lepidoptera: white and sulphur butterflies

R

recurrent veins, 124, 126
root-maggot flies. See *under* Diptera
rostra, 30, 32, 54
rove beetles. See *under* Coleoptera

S

saqakilitaaq. See Lepidoptera
sawflies. *See* Diptera: common sawflies
scabies. *See under* Araneae
scales, 62, 80, 96, 104
Scathophagidae. *See* Diptera: dung flies
scavengers, 25, 39, 44–45, 56
Sciaridae. *See* Diptera: dark-winged fungus gnats
sclerites, 95
scutella, 30, 94
scuta, 94
setae, 62, 140
sheet web spiders. *See under* Araneae
Silphidae. *See* Coleoptera: carrion beetles
Simuliidae. *See* Coleoptera: blackflies
snout moths. *See under* Lepidoptera
spider mites. *See under* Araneae
spinnerets, 97, 136, 138, 140
spiracles, 9, 11, 84
Staphylinidae. *See* Coleoptera: rove beetles
Sternorrhyncha. *See* Hemiptera
stoneflies. *See* Plecoptera
stylets, 30
subscutella, 94
suluqtuq. See Diptera: crane flies
supercool, 13
Symphyta. *See* Hymenoptera: common sawflies
syphons, 41, 62
Syrphidae. *See* Diptera: hoverflies

T

Tabanidae. *See* Diptera: horseflies
Tachinidae. *See* Diptera: tachinid flies
tachinid flies. *See under* Diptera
taqralikisaq. See Lepidoptera
taralikitaaq. See Lepidoptera
tarralikisaaq. See Lepidoptera
Tenthredinidae. *See* Hymenoptera: common sawflies
terrestrial lifestyles
    of aphids, 34
    of arachnids (*general*), 137
    of dark-winged fungus gnats, 73
    of flies (*general*), 56–57
    of non-biting midges, 71
    of rove beetles, 49
    of stoneflies, 23
    of true bugs and homopteran insects, 31–32
    of true spiders, 139
    of water scavenger beetles, 44–45
Tetranychidae. *See* Araneae: spider mites
thoraces (*general*), 8–10
    cephalothoraces, 136, 138, 140, 142, 144, 148
    of balloon flies, 78
    of blackflies, 66
    of blowflies, 90
    of butterflies and moths (*general*), 96–97
    of caddis flies, 116
    of carrion beetles, 46
    of click beetles, 50
    of cockroaches, 24
    of crane flies, 60
    of flies (*general*), 56
    of ladybugs, 52
    of long-legged flies, 80
    of mosquitoes, 62–63
    of no-see-ums, 68
    of owlet moths, 114–115
    of parasitic wasps, 126
    of predaceous diving beetles, 40
    of snout moths, 108
    of stoneflies, 22
    of tachinid flies, 94
    of true bugs and homopterans, 31
    of warble and botflies, 92
tibia, 90
ticks. *See* Araneae: mites and ticks (*general*)
Tipulidae. *See* Diptera: crane flies
Tortricidae. *See* Lepidoptera: leafroller moths
tracheae, 11, 136, 143, 147
Trichoceridae. *See* Diptera: winter crane flies
Trichoptera, 116–117
true bugs. *See* Hemiptera
true spiders. *See* Araneae
tubercles, 84

*tuktu (*var. *tukturjuk*; *tukturruk*; *tuktuujaq*; *tuktuuyak*). *See* Diptera: crane flies
tussock moths. *See under* Lepidoptera
tympana, 108, 110, 112, 114–115

U

uliniu. *See* Diptera: crane flies
umiaqjuk. *See* Diptera: non-biting midges
urogomphi, 40

V

Vacciniina optilete. *See* Lepidoptera: gossamer-winged butterflies
vertebrates, 32, 57
Vespidae. *See* Hymenoptera: social wasps
Vespula austriaca. *See* Hymenoptera: social wasps
Vespula rufa. *See* Hymenoptera: social wasps

W

warble flies. *See under* Diptera
wasps. *See* Hymenoptera (*general*); Hymenoptera: braconid wasps, parasitic wasps, social wasps
water scavenger beetles. *See under* Coleoptera
weevils. *See under* Coleoptera
white and sulphur butterflies. *See under* Lepidoptera
wing buds, 31
wing margin, 124
wingpads, 24
winter crane flies. *See under* Diptera
wolf spiders. *See under* Araneae